NUREG-1789

10 CFR Part 52 Construction Inspection Program Framework Document

U.S. Nuclear Regulatory Commission
Office of Nuclear Reactor Regulation
Washington, DC 20555-0001

AVAILABILITY OF REFERENCE MATERIALS
IN NRC PUBLICATIONS

ABSTRACT

The information contained in the Construction Inspection Program Framework Document, NUREG-1789, details the overall philosophy and approach that the NRC will use to inspect new nuclear power reactors being licensed and built under 10 CFR Part 52. The information detailed in this NUREG about the construction of new reactors will guide the development of inspection manual chapters and inspection procedures that will be used to implement the construction inspection program.

CONTENTS

EXECUTIVE SUMMARY

This framework document will be used as the guiding document for the creation of construction inspection manual chapters and inspection procedures to support the 10 CFR Part 52 licensing process. The staff initially published this document in May 2003 to solicit stakeholder comments. After receiving and considering stakeholder comments, the staff has revised the document and is issuing NUREG-1789 to describe the final construction inspection framework.

In 1991, the Office of Nuclear Reactor Regulation (NRR) started revising the Construction Inspection Program (CIP) governed by Inspection Manual Chapter (IMC) 2512, "Light Water Reactor Inspection Program - Construction Phase." This project had two purposes: to address NRC construction inspection programmatic weaknesses that had been identified during the licensing of several plants, and to develop an inspection program for evolutionary and advanced reactors. The project was stopped in the mid 1990s because no new nuclear power plants were being constructed. A report was assembled that provided an approach reactivating a future construction inspection program. The "Draft Report on the Revised Construction Inspection Program," was issued in October of 1996.

In SECY-01-0188, "Future Licensing and Readiness Assessment (FLIRA)," the NRR staff recognized the need to resume revising the CIP and, in 2001 a CIP team consisting of NRR and regional inspectors was established to do so. A steering committee consisting of NRR and regional managers was also formed.

This CIP framework document updates the work that was published in 1996. It incorporates the applicable requirements of 10 CFR Part 52 such as inspections, tests, analyses, and acceptance criteria (ITAAC), and provides allowances for the rapid construction schedules made possible by the parallel and modular construction techniques used in today's construction environment. Whether or not a combined license (COL) applicant references a certified design, the COL application will contain ITAAC. Should the Commission grant the COL, the staff's construction inspection activities will need to be organized to document the staff's determinations on the licensee's completion of ITAAC.

It is expected, however, that inspections for plants licensed in accordance with 10 CFR Part 52 will be conducted in the same manner as during the construction of earlier reactors, using the same procedures revised, updated, and supplemented by new administrative inspection processes. The Part 52 CIP, therefore, is essentially an updated revision of the older NRC construction inspection program previously used to inspect all light water reactors licensed under 10 CFR Part 50. An electronic information tracking and scheduling system (the CIP Information Management System) is being developed to coordinate inspections with licensee construction schedules and to facilitate access, retrieval, and tracking of inspection findings, reviews, and determinations about the licensee's ITAAC. Appendix A contains a glossary for terms that are used throughout this report.

The program has four phases. The first phase supports a licensing decision for an early site permit (ESP), the second phase supports issuance of a combined license (COL), and the third and fourth phases support construction activities and the preparations for operations.

Inspections will initially be performed to confirm the accuracy of data submitted to the NRC in support of safety evaluations and the Atomic Safety and Licensing Board (ASLB) hearings for an ESP and COL. Because work, such as data collection and procurement, may take place before the staff receives an ESP or COL application, the inspection activities could begin prior to an application. During plant construction, the focus will shift to verifying satisfactory completion of ITAAC, as specified in the final safety analysis report (FSAR), and also to inspecting programs for operational readiness and transition to power operations.

With ITAAC structured as they are, the staff will need to make determinations regarding the completion of individual ITAAC, as the licensee indicates completion of them. Therefore, a phased verification program was developed to assess completion of activities. A sign-as-you-go (SAYGO) methodology will be used, beginning at the early stages of construction; the staff will publish their determinations on individual ITAAC as they are completed; and finally, after all ITAAC have been completed, the Regional Administrator, through the Director of NRR, will make a recommendation to the Commission about whether or not the ITAAC have been met.

Inspectors will verify the completion of ITAAC and document their determinations in inspection reports which will be published on an NRC Web site. At appropriate intervals during construction, ITAAC determinations will be published in the *Federal Register*. It is expected that most negative inspection findings will be resolved primarily by the licensee's corrective action program, but more significant inspection findings may require NRC management involvement. All inspection findings, assessments, ITAAC determinations, and open items will be tracked by the CIP Information Management System.

Since the NRC has limited resources and uses a sampling inspection methodology, reduction in inspection effort may occur when reviews have identified effective program implementation that provides high confidence in the licensee's quality control process. Such reviews will determine whether construction process controls associated with a particular activity are satisfactory. Initially, the activity is heavily inspected. If the process controls are found acceptable, the resources are reduced and that activity is inspected less frequently. Examples of these types of activities are welding, cable pulling, installing pipe supports, and installing electrical penetrations.

The NRC will conduct inspections of operational programs. The scope of the inspections will be similar to the scope of the previous construction program. Most of these NRC inspection findings will be resolved by the licensee's corrective action program, or by traditional enforcement measures. All inspection findings are entered into the NRC CIP Information Management System for easy access, tracking, sorting, and retrieval.

ACKNOWLEDGMENTS

This framework document was developed by a team of individuals from the Nuclear Regulatory Commission's headquarters and regional offices. In addition, a steering group consisting of headquarters managers and a regional manager provided comments and guidance for this document. The construction inspection program steering group is composed of the following managers:

Stuart A. Richards, Chief, Inspection Program Branch, NRR
James E. Lyons, Program Director, New, Research and Test Reactors, NRR
Charles A. Casto, Director, Division of Reactor Safety, Region II

The construction inspection program team that finalized this document is composed of the following members:

Mary Ann Ashley, Inspection Program Branch, NRR
Jerome Blake, Region II
Antone Cerne, Region I
Thomas Foley, Inspection Program Branch, NRR
Ronald Gardner, Region III
Caudle Julian, Region II
Edmund Kleeh, Inspection Program Branch, NRR
Charles Paulk, Region IV
Joseph Sebrosky, New Reactor Licensing Section, NRR

The team would also like to acknowledge the contributions of Michele Evans, Jim Isom, and Carl Konzman during the development of the draft framework document.

ABBREVIATIONS

ABWR	Advanced Boiling Water Reactor
ADAMS	Agencywide Documents Access and Management System
ASLB	Atomic Safety and Licensing Board
CIP	Construction Inspection Program
CIPIMS	Construction Inspection Program Information Management System
COL	Combined License
DAC	Design Acceptance Criteria
ESP	Early Site Permit
FLIRA	Future Licensing Inspection and Readiness Assessment
IMC	Inspection Manual Chapter
ITAAC	Inspections, Tests, Analyses, and Acceptance Criteria
LWA	Limited Work Authorization
NRC	U.S. Nuclear Regulatory Commission
NRR	Office of Nuclear Reactor Regulation
ORAT	Operations Readiness Assessment Team
PM	Project Manager
PRA	Probabilistic Risk Assessment
QA	Quality Assurance
QC	Quality Control
ROP	Reactor Oversight Program
SAYGO	Sign As You Go
SER	Safety Evaluation Report
SSC	System, Structure, or Component
WPC	Work Planning Center

1. INTRODUCTION

1.1 History

In 1991, an NRR working group was established to revise the existing IMC-2512, Light Water Reactor Inspection Program Construction Phase, to incorporate lessons learned from previous construction experience and to develop a program to inspect evolutionary and advanced reactors that might be licensed under 10 CFR Part 52.

At the start of program development in 1991, the working group collated the construction inspection experience within the NRC. The working group revised the inspection programs, policies, and structure and issued a draft report on the revised construction inspection program (CIP). At the same time, the working group developed a computer-based inspection scheduling system to assist the NRC staff in implementing the CIP. The final computer system was based on a system completed for Bellefonte nuclear plant but was also intended for deployment at other nuclear power plants under construction. Because of the lack of new reactor construction activities, the inspection scheduling system was never field-tested. Subsequently, the staff refined the system and called it the CIP Information Management System (CIPIMS).

The CIPIMS provided enhanced guidance and capabilities for gathering, recording, and reporting construction inspection information. The enhancements involved the use of a systems-based inspection planning methodology, the computerization of the inspection program, and a continuous onsite inspection presence throughout plant construction.

The development of the computer program continued until the mid 1990s, when the project was suspended because of NRC staff resource constraints and a lack of nuclear power plant construction. CIPIMS is described in the "Draft Report on the Revised Construction Inspection Program," dated October 1996. In addition to describing CIPIMS, the report presented a framework for the reactivation of a future construction inspection program.

In SECY-01-0188, "Future Licensing and Inspection Readiness Assessment," dated October 12, 2001, the staff recognized the need to restart this effort. The Office of Nuclear Reactor Regulation (NRR) established a CIP team and steering group and began routine public meetings to discuss construction inspection related issues. The CIP team used the 1996 draft report on the revised CIP as a framework to reactivate the dormant CIP.

The team recognized that several assumptions about the construction inspection program have changed since 1996. For example, the 1996 draft report on the revised CIP was written so it could be used either for a 10 CFR Part 50 licensing process or for the newer 10 CFR Part 52 licensing process. While the 10 CFR Part 50 process can still be used to license and construct a nuclear power plant, the industry has indicated that it does not intend to use this process. Additionally, the main focus of the draft report was on the inspection of actual construction activities because that is where most of the work in revising the construction inspection program needed to be done. Early site permit inspection guidance was only given a slight mention in the draft report of 1996. The team recognized that detailed early site permit (ESP) inspection guidance needed to be developed rapidly in order to support the ESP application schedules proposed by industry.

Based on the changes to the CIP which have occurred since 1996, the team decided to update the CIP and to issue new manual chapters for inspections under the 10 CFR Part 52 process. These new inspection manual chapters (IMCs) are based on previous guidance in the IMCs used to assess construction activities for nuclear power plants constructed in accordance with 10 CFR Part 50. The new IMCs also consider and incorporate many of the lessons learned that are discussed in Appendix B of this document.

1.2 10 CFR Part 52 Process

Future U.S. nuclear power plants will be licensed under either 10 CFR Part 50 or 10 CFR Part 52. The new inspection manual chapters have been structured to accommodate the 10 CFR Part 52 licensing process. A brief description of the 10 CFR Part 52 licensing process is provided below in order to place the various construction inspection manual chapters in context:

In 1989, the NRC established new alternatives for nuclear plant licensing under 10 CFR Part 52. Part 52 describes a combined licensing process, an ESP process, and a standard plant design certification process. This approach allows early resolution of safety and environmental issues. The issues resolved by the design certification rulemaking process and during the ESP hearing process are not reconsidered during the COL review except under narrow, clearly defined circumstances. Figure 1.1, "Combined Licenses, Early Site Permits and Standard Design Certifications," below shows the relationship among the three processes.

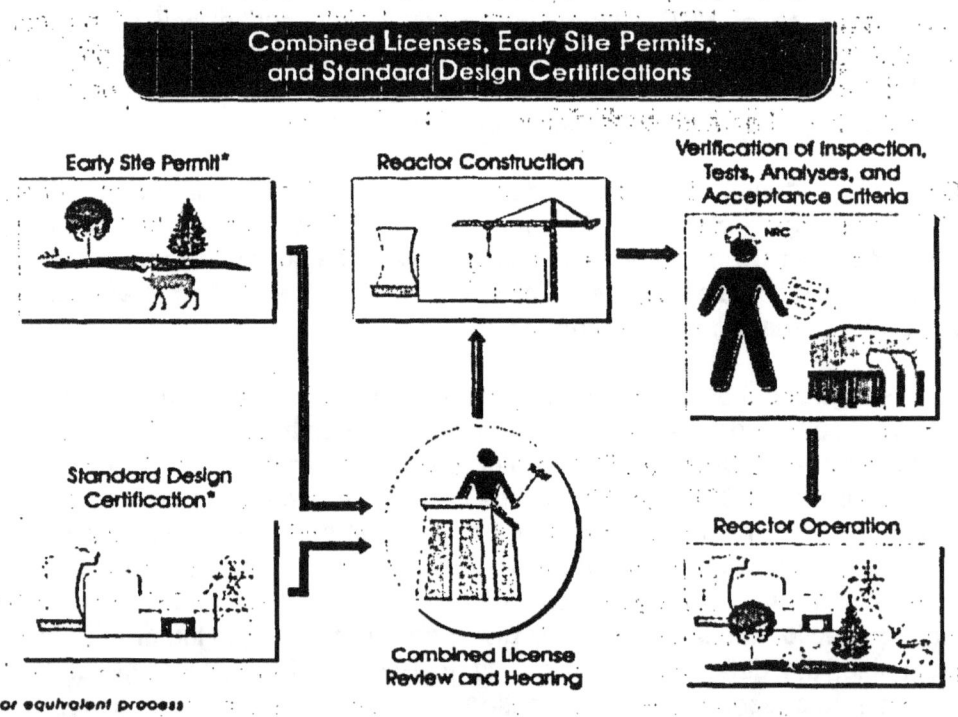

Figure 1.1. Combined Licenses, Early Site Permits and Standard Design Certifications

2

The construction inspection manual chapters provide guidance for the activities in the 10 CFR Part 52 process. The following paragraphs briefly describe ESPs, standard design certifications, and COLs.

- Early Site Permits

Under the NRC's regulations in 10 CFR Part 52, the agency can issue an ESP for approval of one or more sites separate from an application for a construction permit or COL. Such permits are good for ten to twenty years and can be renewed for an additional ten to twenty years. They address site safety issues, environmental protection issues, and plans for coping with emergencies, and are independent of the staff's review of a specific nuclear plant design. Because this is a new process, IMC-2501 and inspection procedures were developed and issued to provide guidance for inspections to be performed to support the issuance of an ESP.

- Standard Design Certification

The NRC can certify a reactor design for fifteen years through the rulemaking process, independent of a specific site. As set forth in 10 CFR 52.47, an application for a standard design certification must contain information describing the design and proposed inspections, tests, analyses, and acceptance criteria (ITAAC) that are necessary and sufficient to provide reasonable assurance that, if the ITAAC are performed and the acceptance criteria met, a plant which references the design is built and will operate in accordance with the design certification.

- Combined License (COL)

A COL authorizes construction and operation of a nuclear power plant with conditions. The application for a COL must contain essentially the same information required in an application for an operating license submitted under 10 CFR Part 50, including financial and antitrust information and an assessment of the need for power. The application must also describe the ITAAC that are necessary to ensure that the plant has been properly constructed and will operate safely.

An application for a COL may reference a standard design certification, an ESP, both, or neither. If the application references a standard design certification, the applicant must perform the ITAAC for the certified design and the site-specific design features. If the application does not reference a standard design certification, the applicant must provide complete design information, including certain information that the applicant would otherwise have submitted for a standard design certification. Similarly, if the application does not reference an ESP, the applicant must provide detailed siting information that would otherwise have been provided during the ESP process.

Should the NRC issue a COL, the NRC then verifies that the licensee has completed the required ITAAC and that the ITAAC acceptance criteria have been met before the plant can operate.

1.3 Expected Licensing and Construction Environment

New certified designs with accelerated construction schedules are being marketed to improve the overall cost effectiveness of nuclear power generation. The accelerated construction schedules are based, in part, on the modular design of these reactors. For the new generation of light water reactors, and for the gas-cooled reactor, the staff understands that applicants plan to have many of the systems/subsystems fabricated at remote facilities (e.g., U.S. or foreign-based shipyards), then to ship these systems/subsystems to the facility for construction of the unit(s). Figure 1.2, "Modular Construction Diagram," provides a representation of this modular construction technique. The staff has had discussions with several different vendors and all plan to use these techniques for plants constructed in the U.S. The nuclear power industry in China and Japan is also currently employing modular construction techniques.

Under 10 CFR Part 52, fabrication activities can begin even before an applicant announces its intent to submit an application for a COL. Therefore, fabrication activities could begin off-site prior to Commission approval of a COL application, but will most likely not begin before an applicant submits an application for a COL.

The expected rapid pace of future nuclear power plant construction will call for the NRC to schedule inspection activities in a way that will ensure that construction inspection does not become a critical path activity. While not required by regulations, an applicant should notify the NRC before fabrication activities begin, to allow the staff sufficient time to plan and implement inspection activities. To assist in more effective inspection scheduling, the applicant's construction plan should be incorporated, if possible, into the construction inspection schedule. Close coordination between inspection and construction schedules will be needed.

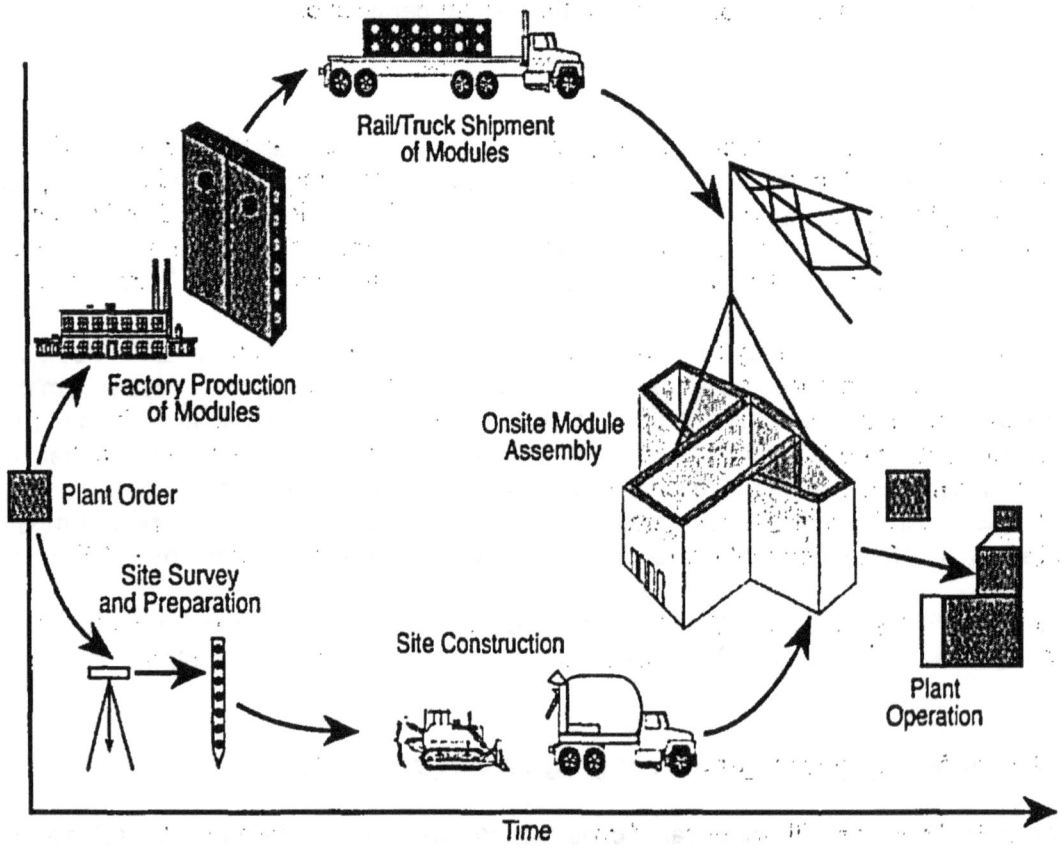

Figure 1.2 - Modular Construction Diagram

4

Depending on the extent of modular construction employed, the inspection staff may need to verify ITAAC at remote facilities during the pre-COL phase. In general, however, critical attributes of systems, structures, and components (SSCs) should be inspected on site to the maximum extent possible. Scheduling inspections at fabrication facilities may be difficult but is important since the fabrication of modules and major plant components could begin many months before the COL is issued and before the first structural concrete is placed.

Evaluations of modules intended to be installed into a plant will be conducted to identify potential modes of degradation during transit.

2. FRAMEWORK OVERVIEW

The Part 52 CIP is essentially an updated revision of the older NRC construction inspection program previously used to inspect all light water reactors built in the United States under 10 CFR Part 50. The old program consisted of five inspection manual chapters. The Part 52 CIP modifies each of these manual chapters to incorporate the applicable requirements of 10 CFR Part 52. The same types of inspections are expected to be conducted for a plant licensed in accordance with Part 52 as those performed during the construction of earlier reactors. These inspections will use the same procedures, as revised, reorganized, updated and supplemented by new administrative inspection processes and the CIPIMS, an electronic information tracking and scheduling system. The CIPIMS was developed to coordinate the inspection and licensee construction schedules and to track inspection findings and ITAAC completion status.

Inspections will focus on two areas: (1) verifying satisfactory completion of ITAAC as specified in the final safety analysis report and (2) compliance with regulations (e.g., Part 21; 10 CFR 50, Appendix B; 10 CFR 50.55(e)), which may not be directly related to an ITAAC. If the inspections identify deficiencies such that an ITAAC was no or will not be met, or that the licensee is not in compliance with regulations the staff will document and docket the information and make it publicly available. The licensee will be expected to take appropriate corrective actions to address deficiencies. The staff may take enforcement action, as appropriate, for instances where regulations have not been met. If the deficiency is not corrected, the Commission may elect not to authorize fuel load. The IMC guidance, for the most part, is written assuming that the licensee will be able to correct such deficiencies.

The staff will verify the completion of certain ITAAC by simply comparing system performance measurements and observations against established criteria. ITAACs of this type will normally be accomplished within a well-defined period during construction and their completion will be easily documented. The licensee will complete the inspection, test, and analysis for other ITAAC, such as welding, over a longer period during construction, and the NRC will perform many inspections to verify their various attributes. When the final construction activity for this type of ITAAC is completed, results of the NRC inspections will contribute to a staff determination of the successful completion of the activity and will ultimately support a Commission finding that all of the ITAAC have been met. In order to allow timely verification of ITAAC that will be done over a long period of time, a "Sign As You Go" (SAYGO) method will be used. This method will require NRC inspectors to sign off completed ITAAC, or portions of complex ITAAC, early in the process as they are successfully demonstrated to the inspector, hence, sign-as-you-go. In accordance with 10 CFR 52.99, at appropriate intervals during construction, the NRC will publish notices of the successful completion of ITAAC in the *Federal Register*. The staff's inspection findings and assessments with respect to ITAAC will be published in inspection reports. This method will provide an on-going record of the acceptability of the work related to the ITAAC.

The new CIP inspection manual chapters can be broken into two categories: (1) Chapters that are done in order to support a licensing decision (i.e., IMC-2501 for an ESP, and IMC-2502 for issuance of a COL), and (2) chapters that are done to verify aspects of construction activities and to provide for the transition to the operations phase (i.e., IMC-2503 and IMC-2504).

The following inspection manual chapters will provide guidance for various inspection activities during the reactor construction period:

(1)   IMC-2501, "Early Site Permit," was issued on October 8, 2002. It is implemented when the NRC is formally notified that an applicant is preparing an application for an ESP. It provides guidance for staff inspection activities from that time, through the receipt of an ESP application, to the subsequent safety review in support of the mandatory hearing that will lead up to the Commission's decision to approve or disapprove the application for an ESP.

(2)   IMC-2502, "Pre-Combined License (Pre-COL) Phase," will provide guidance for inspection activities from the time the NRC is notified of a person's intent to apply for a COL, through the receipt of an application for a COL, to the mandatory hearing that leads to the Commission's decision to approve or disapprove an application for a COL. This IMC and the associated inspection procedures (IPs) will be used to facilitate the inspection activities necessary to support the safety review leading up to the public hearing. The staff will also use this guidance in overseeing any construction activities permitted under § 52.91.

(3)   IMC-2503, "Inspections, Tests, Analyses, and Acceptance Criteria (ITAAC)," will provide guidance for inspection activities to support the staff's review of the licensee's claim that ITAACs have been met. The results of the staff's review will support the Commission's decision on whether to allow fuel loading for a facility that has an approved COL.

(4)   IMC-2504, "Non-ITAAC Inspections," will provide guidance for inspection activities for the requirements of 10 CFR Part 52 and the pre-operational testing phase. Prior to fuel load, IMC-2504 will provide guidance for inspections other than ITAAC. After fuel load, the guidance in this IMC will be used for inspections during initial fuel load, startup and power ascension testing and will be used to guide the transition to IMC-2515.

Inspections associated with IMC-2503 and IMC-2504 will be conducted in parallel and could start at placement of contracts for major component and module manufacturing. However, inspections associated with IMC-2503 end at fuel load while those related to IMC-2504 will end when IMC-2515 is fully implemented.

3. FRAMEWORK IMPLEMENTATION

3.1 IMC-2501, "Early Site Permit"

3.1.1 Introduction

The requirements and procedures for approval of a site or sites for one or more nuclear power facilities are defined in 10 CFR Part 52, Subpart A, "Early Site Permit" (ESP). Inspection Manual Chapter (IMC) 2501 establishes guidance for NRC inspection activities directed towards both pre-ESP application audit activities and post-application inspection activities.

The principal regulatory objective of the ESP phase is to verify that the ESP application meets the requirements specified in 10 CFR Part 52. The application for an ESP must address three areas. First, it must provide a description and safety assessment of the suitability of the site on which the facility is to be located, including the seismic, meteorological, hydrological, and geologic characteristics of the site. Second, it must provide a complete environmental report. Third, it must identify physical characteristics unique to the proposed site that could pose a significant impediment to the development of emergency plans. In addition, the applicant, at its option, may propose major features of the emergency plan or complete an integrated emergency plan for NRC review in consultation with the Federal Emergency Management Agency.

The NRC staff should also explain to the public the contents of an ESP application, the NRC licensing and enforcement process, and the opportunities for public participation. Finally the staff should ascertain whether the elements and standards appropriate to assure quality are being applied to the applicant's ongoing project activities.

Inspections and audits are conducted to verify the quality and accuracy of data collected and the analysis and the evaluation of information used in support of the ESP application.

3.1.2 Audits/Inspections

The ESP phase for a plant is implemented when the NRC receives written notification of an applicant's intention to apply for an ESP under 10 CFR Part 52. The inspection program is applicable to the applicant and the applicant's contractors and to all activities related to NRC regulations.

3.1.2.1 Pre-Application Audits

Before receiving the application, and as soon as possible after being notified of the applicant's intention to submit an ESP, the staff holds meetings with the applicant to establish the primary contacts for the various technical disciplines, review the applicant's schedule for data collection, and arrange to observe the applicant's implementation of its data collection program. The meetings are also used to arrange a preliminary walk-down of the prospective site and to review the applicant's controls for assuring quality in the application. The NRC coordinates schedules with the applicant and gathers information in preparation for public meetings, schedules public meetings to introduce the local community to the NRC licensing process, and arranges meetings with State and local officials.

During the pre-application phase, the NRC conducts audits of the applicant's pre-application activities to check for problems that could lead to an application being rejected. The audits gather information primarily regarding the quality of site suitability data and environmental data collected and the quality of analytical methodologies used in support of the application.

3.1.2.2 Post Application Inspections

During the post-application period, inspections are conducted to support the staff's safety evaluation report testimony for the Atomic Safety and Licensing Board (ASLB) hearing required by 10 CFR 52.21. Based on the information provided by the applicant and the results of the inspections, safety evaluation reports (SERs) are issued and the ASLB hearing is conducted prior to making a determination on whether to grant the ESP.

Inspections are accomplished by the regional office having geographical jurisdiction over the proposed site, with technical support from NRR. Inspections are led by the responsible region after coordinating the effort with the responsible NRR project manager (PM). Technical support is provided by various divisions within NRR as requested by the PM. The technical staff evaluates the applicant's methodologies for data collection. Inspections are consolidated, when practicable, to minimize impact on the applicant. Shortly after the conclusion of each inspection, the NRR technical staff forwards its findings to the inspection team leader for integration into an inspection report.

The inspection procedure guidance in Enclosure 1 to MC-2501 provides the inspector with the applicable inspection procedures for use during inspections, audits, or site visits.

3.1.3 Enforcement

Enforcement actions associated with an ESP application are not anticipated in the pre-application phase. However, as stated in Section 52.21, an ESP is a partial construction permit and is therefore subject to all procedural requirements in 10 CFR Part 2 applicable to construction permits. The information submitted with the application is subject to NRC regulations, including enforcement actions for incomplete or inaccurate information.

3.1.4 Quality Assurance

During ESP activities the applicant should implement QA measures that are equivalent in substance to the requirements of 10 CFR Part 50, Appendix B. This is necessary because the Commission, in proceedings on construction permits, operating licenses, or COLs, treats as resolved those matters resolved during the ESP proceedings, as required by 10 CFR 52.39. Because of this finality, conclusions derived during the ESP phase will be relied upon for use in subsequent design, construction, fabrication, and operation of a reactor that might be constructed on a site for which an ESP has been issued. Therefore, the quality measures implemented for activities important to safety should be equivalent during the ESP and COL phases.

ESP activities associated with site safety assessment should be controlled by QA measures equivalent in substance to the controls described in Appendix B to 10 CFR Part 50. The site safety assessment establishes information, such as analyses and data, that is material to the reliable performance of SSCs important to safety and will be used in the design, construction, and operation of reactor systems that might be constructed on the proposed site. The QA measures provide adequate confidence that SSCs important to safety that are designed and constructed using data and/or analyses derived from ESP activities would perform satisfactorily in service. For example, activities associated with data collection, analysis, and evaluation for soil composition, geology, hydrology, and seismology determinations should be subjected to QA controls, commensurate with the importance of the respective activities to design, and equivalent to the controls described in Appendix B to 10 CFR Part 50. Further, information derived from recognized authorities, such as the Census Bureau or the National Oceanic and Atmospheric Administration, should be controlled using processes for data integrity, data traceability, document control, data evaluation, data analysis, and record storage that are equivalent to the processes and controls described in 10 CFR Part 50, Appendix B.

The pre-application review places particular emphasis on the areas of organization, the QA program, document control, and methodologies for data collection, analysis, and evaluation. It is recognized that certain aspects of the applicant's quality controls may not fully implement the 18 criteria of 10 CFR Part 50, Appendix B, because not all criteria may be applicable to ESP activities. However, the application should provide an adequate basis for evaluation of the acceptability of the information in the application.

3.2 IMC-2502, "Construction Inspection Program: Pre-Combined License (Pre-COL) Phase"

3.2.1 Introduction

This portion of the framework discusses the program of inspections necessary to support the NRC staff's preparation for a mandatory hearing before the ASLB and the final Commission decision on whether a combined license may be granted. This support includes inputs to the safety evaluation report (SER) and public meetings as necessary.

3.2.2 Inspections

The inspections governed by IMC-2502 will be implemented when the NRC is notified in writing that a prospective applicant is preparing to apply for a COL. An application for a COL may, but is not required to, reference a standard design certification, or an early site permit, or both. Therefore, a COL application could include any of the following combinations:

- A standard design certification and an early site permit
- An early site permit and no design certification
- A standard design certification and no early site permit
- No design certification and no early site permit

The guidance contained in IMC-2502 will be flexible to accommodate these various options.

3.2.2.1 Quality Assurance Inspections During the Preparation of the Application

In the past, the NRC conducted meetings with prospective applicants and inspected their activities during the preparation of license applications. The meetings and inspections were to provide assurance to the reviewers about the quality of submittals. The NRC expects to continue this practice through the review of quality controls and the inspection of implementation of those controls.

3.2.2.2 Engineering Design Verifications and First-of-a-Kind Engineering Inspections

In the past, NRC conducted design verification inspections late in the construction process. For the next generation of plants, NRC plans to conduct independent design inspections as early in the process as practical; however, these inspections may continue after the COL is issued. Inspections to support the decision to issue the COL will be conducted under IMC-2502 and will assess the viability and implementation of, and results produced by, the applicant's design engineering process. These inspections will assess the applicant's QA design controls and sample design activities related to the site-specific portions of the plant's design.

Additional programmatic inspections, i.e., to monitor the design change process, may continue under IMC-2504 after the COL is issued.

An inspection program will be developed for the inspection of first-of-a-kind (FOAK) engineering for the lead plant of each certified design. The program will be described in a Inspection Manual Chapter and will be similar to IMC-2530, "Integrated Design Inspection Program."

IMC-2502 reflects the following information on engineering design verification provided to the Commission in SECY-94-294, "Construction Inspection and ITAAC Verification."

- Design descriptions and functional system drawings available for review during the design certification and COL application stages are adequate for licensing reviews and final safety determinations, but not for actual construction or construction inspection activities.

- The NRC will inspect and review the adequacy of licensee design engineering early in a construction project, possibly beginning soon after receipt of a licensing application; first-of-a-kind engineering for the lead plant of each certified design will be assessed during these inspections.

11

- NRC will also assess the effectiveness of the licensee's design change process in maintaining the fidelity of high-level certified design information that is translated into construction drawings.

3.2.2.3 Operational Program Reviews and Inspections

In SECY-02-0067, "Inspection, Tests, Analyses, and Acceptance Criteria (ITAAC) for Operational Programs (Programmatic ITAAC)," the staff provided its recommendation that COLs contain ITAAC for operational programs required by regulations, such as training and emergency planning programs. In response, the Commission's staff requirements memorandum dated September 11, 2002, directed the staff to develop guidelines regarding ITAAC for operational programs and to work with stakeholders to resolve issues associated with the guidelines. The staff provided a response to the SRM in SECY-04-0032, "Programmatic Information Needed for Approval of a COL Without ITAAC." The staff position in the SECY paper involves developing COL review guidance and not inspection guidance.

However, the staff's proposal in SECY-04-0032 discusses the possibility of an applicant submitting implementing procedures at th COL stage to avoid programmatic ITAAC. For operational programs which are submitted as part of the COL application, the NRC staff will conduct its first evaluation by reviewing the bases of and inspecting any implementation of these programs. These inspection activities will be conducted to support the review of the application under IMC-2502. If a program has ITAAC (e.g., emergency planning) inspections will be performed under IMC-2503 to verify the ITAAC. Additionally, SECY-04-0032 recognizes the possibility that inspections of programs that do not have ITAAC will be done after a COL is issued. In such cases, inspections of these programs will be done under IMC-2504.

3.2.2.4 Inspections of Other Activities Completed During the Pre-COL Phase

Changes being considered for 10 CFR Part 52 may allow for partial completion of some ITAAC prior to the issuance of the COL. An example of an ITAAC that could be completed during the licensing review is the ITAAC for control room design, which includes the Design Acceptance Criteria (DAC) for the control room without providing the engineering details. In such cases, the COL would contain ITAAC to verify that the control room has been constructed in accordance with the design.

Two authorizations permitted under § 50.10(e) of the Code of Federal Regulations are informally termed LWAs. These authorizations would also be permitted under § 52.91.

LWA-1: Under § 50.10(e)(1), the Director of Nuclear Reactor Regulation may authorize site preparation work, installation of temporary construction support facilities, excavation for nuclear and non-nuclear facilities, construction of service facilities, and construction of structures, systems, and components which do not prevent or mitigate the consequences of postulated accidents. Sections 52.91(a)(1) and 52.91(a)(2) contain the requirements for permitting LWA-1 activities for a COL. Section 52.25 contain the requirements for permitting LWA-1 activities for an ESP.

LWA-2: Under § 50.10(e)(3), the Director of Nuclear Reactor Regulation may authorize the installation of structural foundations for structures, systems, and components which prevent or mitigate the consequences of postulated accidents. An LWA-2 may be granted if, in addition to the findings described above for an LWA-1, the ASLB determines that there are no unresolved safety issues relating to the work to be authorized that would constitute good cause for withholding authorization.

3.2.3 Construction Inspection Program Information Management System (CIPIMS)

The Construction Inspection Program Information Management System (CIPIMS) should be available for scheduling and recording inspections necessary to support the application review. This will be especially important for the documentation of information related to quality assurance and engineering inspections, LWA activities, and ITAAC completions.

3.2.4 Enforcement

Enforcement actions associated with the application are not anticipated, but are not precluded, during the COL review. However, the information submitted with the application will be subject to NRC regulations, including enforcement actions, for incomplete or inaccurate information. In addition, the Commission has proposed to amend Part 52 to add requirements governing the completeness and accuracy of information submitted by an applicant.

An early site permit referenced by the application for a COL is a license similar to a construction permit issued under 10 CFR Part 50. Therefore, violations of conditions of these licenses during engineering design or LWA activities will be subject to enforcement, including notices of violation, civil penalties and orders.

3.3 IMC-2503, "Construction Inspection Program: Inspections, Tests, Analyses, and Acceptance Criteria"

3.3.1 Introduction

The Commission is required by § 52.97(b)(1) to identify within the combined license the inspections, tests, and analyses that the licensee shall perform, and the acceptance criteria that, if met, are necessary and sufficient to provide reasonable assurance that the facility has been constructed and will be operated in conformity with the license, the provisions of the Atomic Energy Act, and the Commission's rules and regulations. This portion of the inspection program framework document discusses the inspection process used for the NRC's verification of licensee conclusions that the ITAAC of a combined license have been met.

The results of this inspection program will provide input for the Commission's determination, in accordance with § 52.103(g), of whether the ITAAC have been met and whether the licensee is allowed to load fuel. The inspections to verify that ITAAC have been met should begin at the start of placement of contracts for major component and module manufacturing and will end with the Commission's decision on initial fuel load. Inspections of major components may occur before the licensee receives its COL.

3.3.2 ITAAC Inspection Overview

Figure 3.1, "IMC-2503, Inspection Flow Diagram," depicts the basics of the inspection process for this phase and the various types of inspection findings/assessments that will be made. Individual ITAAC determinations of acceptability are made after inspection by NRC inspectors and documented in inspection reports, as discussed below. It should be noted that there is only one overall ITAAC conclusion made by the Commission relative to § 52.103(g). To support the §52.103(g) finding, the staff intends to use a phased verification method. This phased verification method includes the concept of planning the inspection effort at the beginning of the process, using a sign-as-you-go (SAYGO) throughout construction, publishing the results of NRC staff determinations with respect to individual ITAAC as they are completed, and finally, making a recommendation to the responsible regional administrator regarding the completion of all the ITAAC.

The NRC staff expects that when an ITAAC is complete, the licensee will provide an ITAAC determination letter that will demonstrate the satisfactory completion of the smallest increment of an ITAAC. Examples are shown in an NEI letter to the NRC dated November 20, 2001 (ADAMS accession number ML 020070338) and include, "ITAAC 2.4.2 Item 7 High Pressure Core Flooder System Remote Shutdown System Display," or ITAAC 2.4.2.2, "High Pressure Core Flooder System Hydrostatic Test." However, there are various possibilities of ITAAC determination letters for the identification of ITAAC completion including: an entire table (e.g., Table 2.1.1.d - Reactor Pressure Vessel); a complete line item from a table (e.g., Table 2.1.2 - Nuclear Boiler System, Items 9.a. and 9.b.); or an individual item (e.g., Table 2.1.2, Item 9.a.). It is important that the licensee schedule the performance of each ITAAC and communicate the schedule to the NRC in order to allow NRC inspectors the opportunity to witness the performance of the ITAAC.

The master NRC construction inspection schedule will be derived from CIPIMS. CIPIMS will contain all the ITAAC, the licensee's construction schedule and all related inspection resource information. This system will also integrate all inspection results and correlate them with ITAAC and non-ITAAC requirements and acceptance criteria. It will facilitate inspection at remote locations as licensees make use of modular construction techniques, and integrate the inspection planning process with the licensee's detailed construction scheduling process. It will allow the NRC to integrate the results of inspection findings, ITAAC determinations, and the staff's evaluation of licensee quality assurance/quality control (QA/QC) effectiveness.

The ITAAC inspections will lead to two possible results: (1) inspection findings, and (2) ITAAC determinations. An example of an inspection finding would be documentation of the staff's evaluation of the acceptability of licensee work processes that affect multiple ITAAC. A second example of an inspection finding would be documentation of the staff's evaluation of a component associated with a particular ITAAC. All inspection results will be documented in inspection reports.

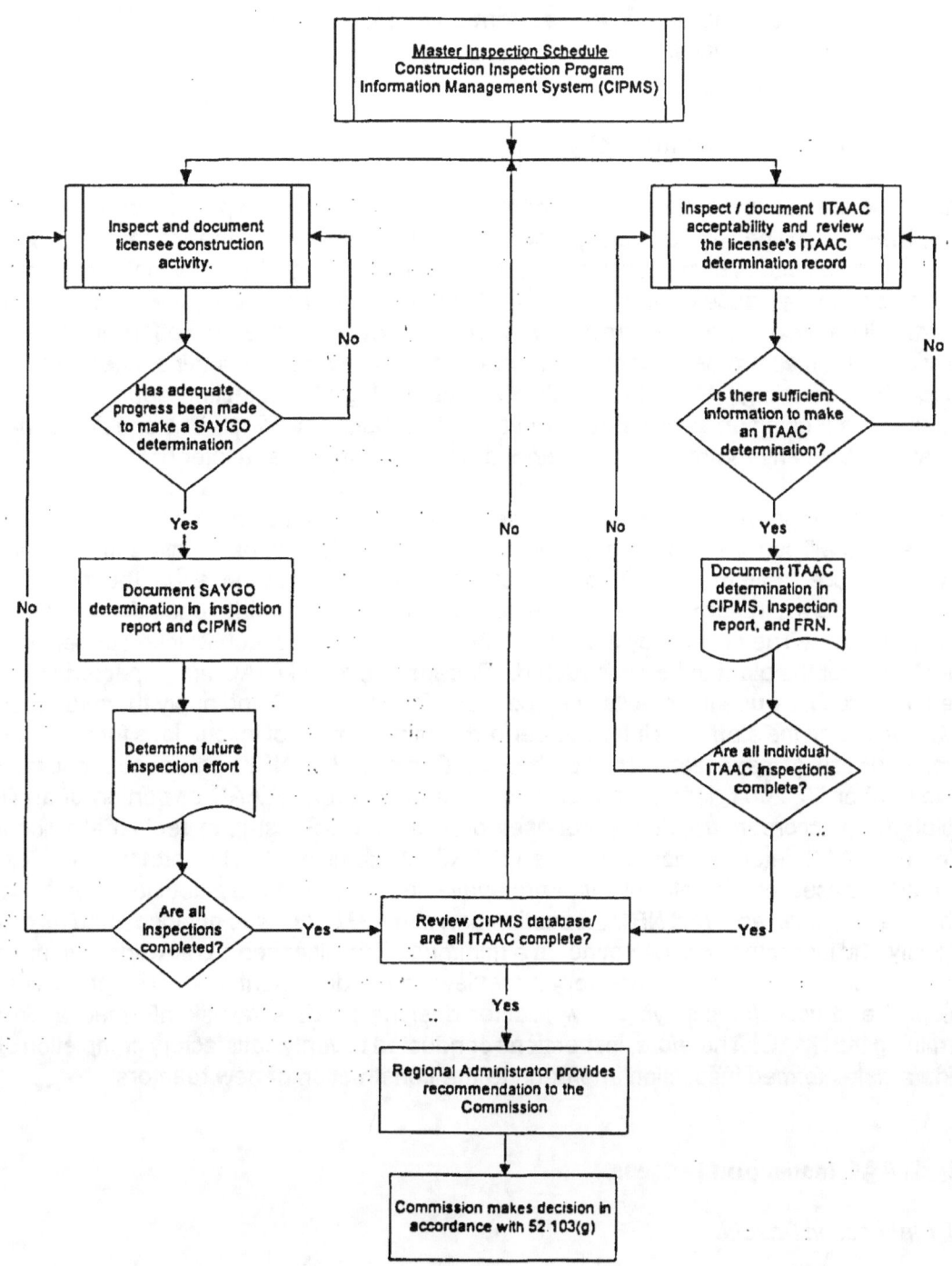

Figure 3.1 IMC- 2503
Inspection Flow Diagram

An ITAAC determination will document the staff's position on whether or not the licensee has satisfactorily demonstrated that a particular ITAAC has been met (i.e., individual items in the ITAAC tables). The regional administrator will be informed periodically on the status of ITAAC inspections and all ITAAC determinations.

3.3.3 ITAAC Inspection Philosophy

Because the staff does not have the resources to perform direct inspection of all elements of all ITAAC, the NRC will perform sampling-type inspections to verify that the licensee is in compliance with NRC regulations. A combination of ITAAC sample selection, statistical methods, insights from the probabilistic risk assessment (PRA), and inspections of the licensee's quality assurance program will be used to help determine the necessary level of inspection effort and where limited inspection resources are best spent. Appendix C of this document contains a general discussion on inspection sampling. Work to establish a methodology for selecting appropriate inspection samples was still going on at the time this document was issued as final. The specific guidance and information about any methodology developed will be published at a later date.

The inspection program will rely on the licensee to ensure that all of the ITAAC have been met and the inspectors will perform sampling type inspections to verify that the licensee has completed the ITAAC in an acceptable manner. This will provide reasonable assurance that the facility has been built and will operate in accordance with the license and the applicable regulations. The sampling type inspections will be planned by the staff at the earliest stages of construction based on a review of the ITAAC for the plant to be constructed. Because several ITAAC are expected to be closely related, the staff may use the results of inspections for one ITAAC and apply them to other related ITAAC. However, the staff does intend to perform a minimum set of inspections for all of the ITAAC. The minimum set of inspections for all of the ITAAC is based on NEI's proposed process, set forth in a November 20, 2001, letter, for informing the staff when an ITAAC or portion of an ITAAC is completed. In accordance with this proposed process, the NRC staff expects that a licensee will provide an ITAAC determination letter when ITAAC are completed. This letter will also inform the staff that the bases for the determination are available for audit at the plant site. For those ITAAC which have not received direct NRC inspection or a similar ITAAC was not inspected as discussed previously, the inspectors will determine, at a minimum, if the licensee's ITAAC determination letter and its associated bases are satisfactory by reviewing the documentation. The process that was developed and used for certifying new reactor designs provided a risk-informed approach for determining the ITAAC. Therefore, inspections conducted to verify satisfactory completion of ITAAC provide a risk-informed inspection approach to the construction of new reactors.

3.3.4 ITAAC Inspection Process

3.3.4.1 Inspection Results

Inspections of ITAAC-related activities will be conducted in accordance with the inspection procedures listed in IMC-2503. Inspection results will be documented in accordance with IMC-0613, "Power Reactor Construction Inspection Reports." The staff has not yet developed IMC-0613, but intends to do so within the next two years.

3.3.4.2 Review of Inspection Results

The review of inspection results will focus on two things: (1) the implementation of specific activities as documented in the inspection history and (2) the implementation of the licensee quality controls. The review would ensure that any deficiencies that have been identified by the licensee or NRC have been adequately addressed by the licensee's QA program and have resulted in effective corrective actions. This would provide the NRC with confidence in relying on the licensee's quality assurance program in assuring quality construction activities.

In keeping with the "sign-as-you-go" (SAYGO) approach, NRC staff and management will periodically review inspection results to determine if the inspection history shows that sufficient progress has been made in a specific area to reach an overall determination of acceptability. Based on the inspection history, the staff could "sign-off" on the activities that have been found acceptable and the level of inspection effort could be adjusted.

(1) Positive SAYGO Determinations

Should a review of the inspection history identify that activities are being effectively implemented and deficiencies are being appropriately addressed, NRC staff will document their review as a positive SAYGO determination. The inspection efforts associated with the particular construction activity or with a specific process may be reduced based on a positive SAYGO determination. This determination could also reduce the inspection effort in other areas which are affected by this same activity or process. The positive determination could also be used by the staff at a later time when making the determination that ITAAC have been met by the licensee.

(2) Negative SAYGO Determinations

If the review of the inspection history identifies that a construction activity is not being effectively implemented and that significant deficiencies are not being identified and appropriately corrected, it could call into question the effectiveness of the licensee's quality assurance program and, if not corrected, prevent the staff from making a positive ITAAC determination. NRC staff will document their review as a negative SAYGO determination.

A negative SAYGO determination would call for the licensee to identify correction actions taken or planned to address the specific identified deficiencies as well as the deficiencies in the corrective action program. Further, the licensee would be expected to determine how the deficiency occurred, whether or not it was generic, and to take actions to determine the extent of the condition.

NRC would consider increasing its inspection effort in this area by expanding the inspection sample size to verify the extent of the condition and, if appropriate, re-examining other ITAAC which may have the same or similar deficiencies. This could also increase the inspection effort in other areas which are affected by this same activity or process. The NRC will verify the effectiveness of any corrective actions. Upon verification of effective corrective actions, NRC staff would reassess the construction activity, process, or component.

All reviews of inspection results would be documented in inspection reports and also reflected in the CIPIMS database.

Table 3.1, "Examples of Construction Processes Appropriate for Evaluation Using a SAYGO Approach," contains examples of construction processes which might be candidates for evaluation using SAYGO. The list is intended to be representative rather than all-inclusive. Other construction processes may also be appropriate and actual construction approaches may make the use of a SAYGO approach inappropriate for some processes at some sites.

If, subsequent to the identification and documentation of a review of inspection findings, the NRC determines that the results are no longer valid (e.g., the NRC, licensee, or any other person identifies new and significant information that has not been adequately addressed by the licensee's corrective action program), the determination would be reassessed by cognizant NRC management, communicated to the licensee, and documented in an NRC inspection report. Consistent with past practices, the licensee would be afforded an opportunity to provide any new information to the NRC which might affect the reversal of the previous review determination.

Table 3.1, Examples of Construction Processes Appropriate for Evaluation Using a SAYGO Approach		
Site preparation	Concrete expansion anchors	Heating, ventilation and air conditioning
Mechanical penetrations	Structural steel and supports	Conduit/tray supports
Equipment fabrication	Safety related piping	Conduit installation
Geotech/foundations	Pipe support and restraints	Tray installation
Structural concrete placement	Welding	Cable pulling
Re-bar installation	Masonry	Cable terminations
Instrument sensing line installation and piping	Mechanical component/ equipment installation	Electrical component/ equipment installation
Nondestructive examination	Electrical penetrations	

3.3.4.3 ITAAC Determinations

As specific construction activities are completed, the licensee will determine that one or more ITAAC have been completed. The licensee will document the specific inspections, tests, or analyses relied upon in making the determination that one or more ITAAC are complete and ready for NRC verification. This will be communicated to the NRC in the form of an ITAAC determination letter requesting that the NRC staff verify that the ITAAC have been satisfactorily completed.

(1) Determining ITAAC Acceptability

Upon receipt of an ITAAC determination letter, the NRC will review the licensee's ITAAC documentation and any NRC inspection reports related to that ITAAC. An NRC staff decision on ITAAC acceptability will be called an ITAAC determination. The NRC's determination of ITAAC acceptability will be based primarily on prior day-to-day onsite and offsite inspection activities, interactions with licensee personnel, and inspection of construction activities in the field. These inspections will have been documented in inspection reports.

In accordance with 10 CFR 52.99, the NRC will document each ITAAC determination in a *Federal Register* Notice and in docketed correspondence to the licensee. The basis for determining the acceptability of ITAAC will be documented in inspection reports and tracked in CIPIMS.

(2) Invalidation of Previously Accepted ITAAC Determinations

If new and significant information questions the validity of a previously accepted ITAAC determination, the NRC would assess the information and determine the appropriate course of action. The threshold that the NRC will use to determine what constitutes "new and significant information" that would invalidate a previous ITAAC determination is illustrated by examples in Appendix D of this framework document.

Consistent with past practices, the licensee would be afforded an opportunity to provide any new information (potentially including extensive corrective actions) to the NRC which might affect the reversal of a previously accepted ITAAC determination. This information would be expected to address whether or not the deficiency could be generic to other ITAAC and also why the extent of the condition is or is not limited to this particular ITAAC. In addition, the licensee would be expected to identify and correct the weaknesses in its corrective action program that allowed the deficiency to occur. The NRC staff's decision on whether the ITAAC has been met would be communicated to the licensee and would be made publicly available via the *Federal Register*.

3.3.4.4 Commission 10 CFR 52.103(g) ITAAC Finding

Before a facility may operate, the Commission is required by § 52.103(g) to find that the acceptance criteria in the COL were met. Once the licensee has informed the staff that all the ITAAC have been completed, the staff will perform a review to ensure that an ITAAC determination letter has been received for each ITAAC, a notice has been published in the *Federal Register* for each ITAAC that the staff has accepted, and the staff agrees that all the ITAAC have been met. The RA will rely on the inspection and ITAAC determination results when informing the Director of Nuclear Reactor Regulation (NRR) that all the ITAAC have been met. The Director of NRR will make a recommendation to the Commission that the Commission find that all acceptance criteria in the COL have been met.

3.3.4.5 Public Notifications

The requirements for public notification of ITAAC completion are contained in § 52.99. The staff intends to publish ITAAC determinations in the *Federal Register.* In addition, NRC inspection reports, and correspondence with the licensee will be published and be made available to the public. The staff is considering the use of the NRC web site, similar to how it is used for the reactor oversight program (ROP), as the chief electronic medium through which the results of inspection activities can be made more readily available to the public.

3.3.4.6 Enforcement

During the construction period, the agency will process identified violations of NRC regulations and conditions of the COL as set forth in the Commission's Enforcement Policy, NUREG-1600, "General Statement of Policy and Procedures for NRC Enforcement Actions," and will track these violations in CIPIMS.

3.3.5 Inspecting Module Construction Activities

The use of a modular construction concept may be necessary to support the ambitious schedules currently being proposed for the construction of the new generation of nuclear power plants. Offsite fabrication of plant modules and plant components could begin well before COL issuance. There may be some instances where a basically complete plant is fabricated at one offsite facility. In such cases, the requirements to have a manufacturing license in accordance with 10 CFR Part 52 may apply.

Major plant components such as reactor pressure vessels, steam generators, and reactor coolant pumps, as well as smaller components such as electrical breakers, relays, and valves, have traditionally been fabricated at an offsite location. For future nuclear power plants, large portions of the plant could be modular in design, allowing for offsite fabrication and assembly of portions of buildings and rooms containing completed and tested systems and subsystems.

Discussions with several design and construction organizations concerning modular construction have convinced the staff that as much as 60 percent of what had been site construction activities in the past will probably be moved offsite to the locations where the modules will be fabricated. NRC oversight activities in addition to those performed during the construction of the existing fleet of nuclear plants will be necessary to assure that an acceptable level of quality is maintained throughout the fabrication or manufacturing process. Collectively, these offsite construction activities pose significant challenges for the planning and implementation of NRC inspections.

One example of such a challenge arose during the construction of the ABWR in Taiwan. On that project, a problem occurred during the fabrication of the reactor pressure vessel pedestal. The problem involved offsite fabrication as well as onsite construction by the fabrication contractor. The staff believes that this problem provides valuable "lessons learned" for the inspection of fabrication facilities and remote process activities. Appendix E to this document contains information regarding the problem.

For all offsite manufacturing and fabrication activities, a major focus of NRC review will be assuring acceptable licensee QA oversight. In addition, for ITAAC-related components and modules, the NRC may perform inspections at the offsite location. A major focus of these offsite location inspections will be assuring that the vendor has implemented QA requirements appropriately. Additionally, the staff intends to inspect the programmatic implementation of QA requirements at vendor facilities and will assess the overall effectiveness of vendor QA activities under IMC-2504. Appendix C of this document provides a more detailed discussion of the role of the quality assurance program as it related to ITAAC. As discussed in Appendix C, QA deficiencies may impact the NRC's ITAAC determinations.

The NRC would expect to use the above-mentioned phased verification process and apply it to offsite inspections where appropriate. For example, if safety-related pipe welding is taking place in several different offsite fabrication facilities (e.g., shipyards) because of modular construction, and also onsite (to connect one safety-related module to another), then inspection findings and assessments could be used for the individual offsite fabrication facilities as well as for onsite activities.

Similarly, an inspection finding or assessment could be used at manufacturing facilities that are supplying the reactor pressure vessel, steam generators, etc. If a positive SAYGO determination is documented, site inspections could be limited to inspections for handling and shipping damage after the component arrives at the site.

Figure 3.2, "Anticipated Nuclear Power Plant Construction Schedule," lays out a typical construction schedule for a nuclear power plant using modular construction techniques. A gas-cooled reactor vendor indicated that the time from first placement of structural concrete to fuel load (all ITAAC met) was projected to be approximately 20 months per module, while another light water reactor applicant indicated that the construction time frame for its design would be 42 months. The 36-month time frame in the figure is, therefore, meant to be representative and the schedules will be different based on the design and the applicant. The top of the figure shows the applicant's schedule and at the bottom of the figure are the staff's two high-level IMCs that will guide the inspection activities associated with this schedule.

The time line for IMC-2503 shows the major milestones associated with the staff's ITAAC inspection activities. These inspection activities would start with inspections associated with an overview of the applicant's QA program and how the applicant will satisfy the QA requirements associated with the fabrication of the major components and modules. The process ends with the Commission decision regarding ITAAC. This time line shows when the staff expects to perform inspections that will have a direct effect on ITAAC determinations. With the heavy reliance on modular construction, the staff fully expects to be performing inspections both onsite and at offsite facilities to support ITAAC determinations.

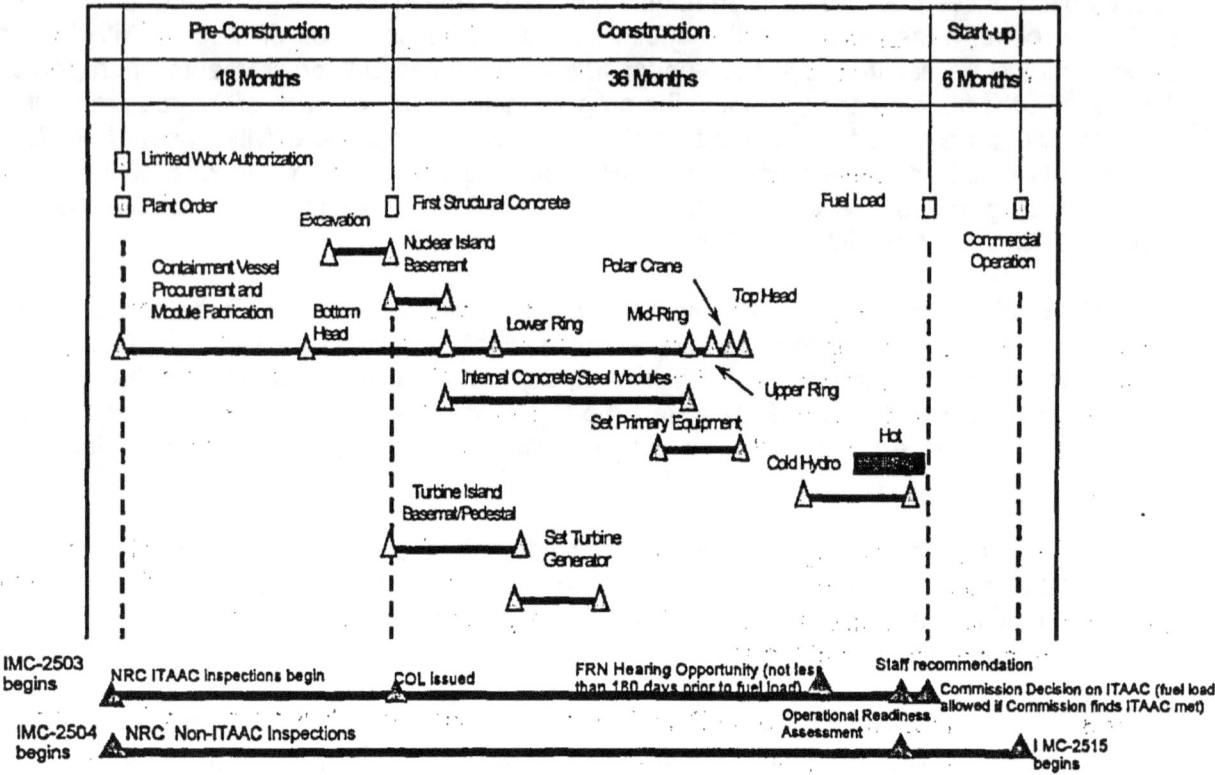

Figure 3.2 - Anticipated Nuclear Power Plant Construction Schedule

The second time line is for IMC-2504, which is discussed in more detail in Section 3.4 of this document. This time line is meant to illustrate the staff's inspections associated with things other than ITAAC, the transition to IMC-2515, and the reactor oversight program.

3.4 IMC-2504, "Construction Inspection Program: Non-ITAAC Inspections"

3.4.1 Introduction

The purpose of this section of the framework document is to discuss the inspections, other than those associated with ITAAC, that the NRC expects to perform from the time that a COL is issued (or earlier if necessary) until sometime after the plant reaches full power operations status, i.e., when IMC-2515 begins. Inspection guidance for a plant licensed in accordance with Part 50 that is relevant to a plant licensed in accordance with Part 52 is contained in IMC-251 and IMC-2514. The scope of inspections will include structures, systems and components that are safety related to ensure that the licensee is in compliance with appropriate rules. In addition, other issues that are not safety related but are considered important to safety, such as station black out, will also be inspected.

Because of the different issues involved with this period, the time frame is broken into two segments; the time period before the § 52.103(g) finding is made by the Commission, and the time period from the §52.103(g) finding until the point when IMC-2515 begins. This division is needed because, at the time of the §52.103(g) Commission finding, several things occur that affect the inspection program. For example, ITAAC end, while emergency planning, and technical specifications requirements begin. Inspections associated with each time period are discussed in the sections below.

3.4.2   Before 10 CFR 52.103(g) Finding

If an operational program has an ITAAC associated with it, then IMC-2503, discussed above, would govern inspections associated with that program. The staff will perform inspections of programs early in construction and prior to operation to verify the licensee's compliance with regulations. The Commission's SRM associated with programmatic ITAAC, dated September 11, 2002, recognized that because not all operational programs will have an ITAAC, the staff can take appropriate enforcement action to prohibit or delay fuel loading pending appropriate corrective action if the licensee's operational programs do not provide adequate protection of public health and safety.

Inspections of programs that do not have an ITAAC would be done similarly to the construction inspection under 10 CFR Part 50. The inspection guidance for operational programs is contained in existing inspection procedures which will be implemented by IMC-2504. The guidance previously contained in IMC-2513 for pre-operational testing under 10 CFR Part 50 will be edited and incorporated into IMC-2504. The staff expects to augment such inspection guidance with lessons learned from the ROP. This may include using some of the inspection procedures usually implemented under IMC-2515 to assess some operational programs and doing inspections in the cornerstone inspection areas identified in the ROP. The cornerstone inspection areas are reactor safety, radiation safety, and safeguards, while the cross cutting elements are human performance, safety conscious work environment, and the corrective action program. Inspections done in these areas are separate from the ITAAC determinations. To the extent that they are performed prior to loading fuel, these inspections will supplement the bases for the regional administrator's recommendation to the Director of NRR regarding the licensee's overall readiness to load fuel.

Engineering design inspections will continue as the licensee completes site-specific designs to address the design acceptance criteria (DAC) and to document the final as-built system configuration.

3.4.2.1 Inspection After an ITAAC is Met and Prior to Fuel Loading

There is guidance in both draft SRP 14.3 and in SECY papers to the Commission on performing inspections after an ITAAC determination is made but before the Commission makes its determination in accordance with § 52.103(g). Draft SRP section 14.3, "ITAAC - Design Certification," provides some guidance in this area. The following is an excerpt from Section 14.3, Appendix A, Section IV.B.2 of the SRP:

> The purpose of the ITAAC is to verify that an as-built facility conforms to the approved plant design and applicable regulations. When coupled in a COL with the ITAAC for site-specific portions of the design, they constitute the verification activities for a facility that must be successfully met prior to fuel load. If the licensee demonstrates that the ITAAC are met and the staff agrees that they are successfully met, then the licensee will be permitted to load fuel. Once completion of ITAAC and the supporting design information demonstrate that the facility has been properly constructed, it then becomes the function of existing programs such as the technical specifications, the in-service inspection and in-service testing program, the QA program, and the maintenance program, to demonstrate that the facility continues to operate in accordance with the certified design and the license.

The staff has also provided guidance on when programs such as technical specifications are required to be in place. Most recently, the proposed rule for Part 52 contains a revised section on applicability of NRC requirements (see section 52.215(c) on page 122 of the attachment to SECY-02-0077). This section makes it clear that the COL holder does not need to be in conformance with most operational requirements until the Commission has authorized fuel load. The staff also proposed in SECY-00-0092, "Combined License Review Process," license condition 2.I (see Appendix F) which states:

> The following operational requirements that are applicable to this license will become effective after the Commission finds that the acceptance criteria in this license (COL ITAAC) have been met in accordance with 10 CFR 52.103(g):
>
> - emergency plans
> - technical specifications

Therefore, there is the expectation that a license condition will specifically designate when some operational programs will be required to be in place.

Because of the above, the staff recognizes that there is a period of time between when an individual ITAAC is completed and when the Commission's § 52.103(g) finding is made. Thus, there will be a period of time from completion of an individual ITAAC to when the programs, such as technical specifications, are required to be in place to demonstrate that the facility continues to operate in accordance with the certified design and license. During this period of time, the NRC will perform inspections in accordance with IMC-2503 to verify that the ITAAC determinations remain valid.

3.4.2.2 Operational Readiness Assessments

In the past, an operational readiness inspection was done in accordance with Inspection Procedure 93806, "Operational Readiness Assessment Team Inspection (ORAT)," to ensure that a plant was prepared for its low power license. Operational readiness inspections will continue to be performed

24

for plants licensed in accordance with 10 CFR Part 52. The results of the inspections could provide significant information separate from the ITAAC completion status to the regional administrator and Director of NRR regarding the licensee's operational capability and organizational readiness to load fuel.

Inspections to observe the development of programs necessary to support major activities in construction and turnover to operation are expected to begin early in the construction process. The staff expects the licensee to phase in the operational programs necessary to support each milestone of construction before the program is required by regulations. These inspections would verify that the licensee is ready to implement programs such as licensed operator training, security and fire protection, which would be necessary to support fuel load. The staff expects to inform the Commission of the status of these programs before a Commission decision is made relative to §52.103(g).

3.4.3 Post Fuel Load Prior to Power Operations

As discussed earlier, ITAAC end when the Commission makes the findings required under § 52.103 before operation. Therefore, after fuel load, the ITAAC do not constitute regulatory requirements for the COL holder. Adequate protection of the public health and safety during plant operation is assured by continuing compliance with the terms of the COL, including technical specifications, and the NRC's regulations. The inspection guidance for this phase of the construction inspection program will be contained in IMC-2504. In addition, as discussed above, lessons learned from the ROP will be implemented so that the transition to IMC-2515 is smooth. In general, IMC-2504 will support the remainder of the pre-operational testing (e.g., integrated hot functional test and loss of offsite power), the startup testing (pre-criticality tests, low power physics testing, and power ascension testing), and transition to IMC-2515.

The staff will continue to perform inspections to support major milestones after fuel loading and before IMC-2515 begins. Such milestones include low power and full power operations. The staff expects that the RA will inform the Director of NRR on the licensee's readiness to achieve these milestones.

3.5 Construction Inspection Program Information Management System (CIPIMS)

CIPIMS is a dedicated, computer-based inspection scheduling and information management system intended for deployment at nuclear power plants (NPPs) under construction. CIPIMS will be used to organize and manage inspection information and will integrate the licensee's construction schedule, inspection results and findings to support ITAAC determinations. CIPIMS will provide a standard, consistent, systems-based approach to coordinating, scheduling, collecting, organizing, and recording inspection data necessary to establish a reasonable assurance finding for ITAAC determinations and eventual transition activities from construction to operational inspection under IMC-2515.

The CIP team and NRR's Work Planning Center (WPC) initiated an effort to adapt the CIPIMS software described in Attachment 4 of the "Draft Report on the Revised Construction Inspection Program," dated October 1996, to the CIP provisions of today. In addition to updating the software, the team also faced the challenges of interfacing the system with the Agencywide Documents Access and Management System (ADAMS), which was created after the original CIPIMS was developed, and applying new technologies that will support CIP efforts to economize inspection resources.

The combined efforts of the CIP team and NRR's WPC resulted in redefining the needs to be met by CIPIMS to accommodate future construction techniques, quality processes, inspection program management, skills and experience, inspection program structure and implementation, inspection documentation, and inspection planning and activities.

CIPIMS will use:
- Smart coding
- Integrated scheduling
- Tablet personal computers
- Barcode technologies for data collection and tracking
- Computer aided engineering (CAE) design tools
- Digital imaging
- Lessons learned from international programs and feedback on foreign parts fabrication
- CIPIMS should have native compatibility with licensee information systems and technologies (licensee Primavera® P3 scheduling systems, Microsoft® Office, etc.).

The CIPIMS scheduling software should be able to easily interface with the licensee's scheduling software. The vision for CIPIMS is that the NRC inspection scheduler, working with the respective region and headquarters, will plan construction inspection activities in advance based on the initial schedule from the licensee. The schedule would then be automatically updated as the licensee's schedule changes. The licensee may not know the inspection activities the NRC has planned in advance but it is vitally important that the staff have up-to-date accurate information regarding the construction schedule, including those activities that are being performed offsite.

A series of meetings with Westinghouse, General Electric, Atomic Energy of Canada Limited, and Bechtel to discuss their construction scheduling software revealed that each is currently using Primavera® as their scheduling software program. However, they acknowledge they may eventually move to other products. The NRC has noted that the level of detail in some master schedules, where only the delivery date of a major component or module is identified, may make scheduling an inspection more difficult.

The NRC and NEI have established a working group specifically to test CIPIMS. The group will establish the level of detail, coding structure, and transfer protocols needed to efficiently transfer schedule information for use in CIPIMS. These outcomes will be achieved by developing and transferring detailed schedules on selected sample work streams as a means of testing the various attributes of CIPIMS.

CIPIMS will be successful only if the licensee, the prime, sub, and fabrication contractors, and the NRC use a common coding schema allowing seamless integration of licensee schedules with

ITAAC procedures. The adoption of a standard coding schema will provide for auditing and traceability of the inspection process at a level of detail previously not available to the NRC. It should be noted that if fabrication occurs in geographically dispersed areas, support for universal time and date formats will be necessary to ensure proper resource scheduling and availability.

3.6 NRC Organization

3.6.1 Introduction

Upon receipt of a written notice of intent to submit an application for an ESP or a COL, NRC management will establish an NRC organization to implement the CIP for the proposed site.

3.6.2 Implementation

The CIP team determined that although the specific organizational composition would be best defined at the time of implementation, there were some important factors to be considered when initially establishing this NRC organization.

3.6.2.1 ESP

The area of inspection is fairly limited for an ESP application, and therefore major organizational considerations are not involved. It is important that NRR and the Region identify points-of-contact (i.e., a project manager or a project engineer) as soon as an applicant announces the intent to submit an ESP application. These points-of-contact are necessary for the coordination of pre-application site visits, review inspections, and public meetings.

3.6.2.2 COL and Beyond

Inspections to support the review of a COL application are more involved than inspections for an ESP. Once again, it is important that NRR and the Region identify points-of-contact early (i.e., as soon as the intent to submit a COL is announced) in order to coordinate necessary pre-application activities. One important function of the pre-application contacts with an applicant is to gauge the applicant's proposed construction schedule and projected offsite construction plans, in order to determine when the NRC should begin organizing its construction inspection team.

3.6.3 Items To Be Considered During the Development of the Inspection Organization

The licensing and construction environment needs to be considered when establishing the organization. During discussions with design organizations, the CIP team was informed that by using modular construction techniques, as much as 60 percent of what in the past were considered typical onsite construction activities will be moved off site. It is also conceivable that some offsite modular and component fabrication activities could begin prior to the submittals of the application for a COL.

Depending on the extent of modular construction employed, the inspection staff may need to conduct inspections at remote locations, such as U.S. or foreign shipyards or fabrication facilities. Because of modular construction, overall site construction schedules will be significantly accelerated compared to those of past construction projects. Site construction duration from the first safety-related concrete placement until request for permission to load fuel could be as short as 24 to 30 months.

Another consideration which would have an impact on the development and location of an inspection organization is whether one COL application will be submitted or whether multiple applications are expected.

Considering these factors, the CIP team identified the following as necessary to ensure the success of the initial organization.

3.6.3.1 Regional Involvement

The regional office should oversee the implementation of, and the overall coordination of, the inspection program for a particular site. This is necessary to ensure that the cognizant regional administrator is involved throughout the process so that the RA may make a recommendation to the Director of NRR regarding the completion of ITAAC and other prerequisites for initial fuel load.

The regional office should oversee the onsite inspection program and would provide inspection resources and other technical support as necessary.

3.6.3.2 Inspection Scheduling and Data Management Activities

There will be a critical need for a central scheduler. There is a likelihood that the CIP implementation could be performed using one organization to address onsite construction and a different organization to address remote fabrication of components and modules. Inspection results from separate inspection locations could be used to complete a single ITAAC.

In order to plan for and to coordinate all of the required inspection activities, the scheduler would be directly responsible for communicating with the applicant's scheduling organization and for coordinating all inspection activities with the respective inspection team leaders. All inspection activities coordinated through the central scheduler would need to be planned, scheduled, and tracked through completion using a central data system such as CIPIMS.

The scheduler would be trained in a scheduling program that is compatible with the applicant's scheduling software. This individual would also be trained on the use of the CIPIMS and would be responsible for the overall utilization and maintenance of the CIPIMS data for that site.

3.6.3.3 Inspection Project Management

In the earliest phases of plant construction, inspection activities could be governed by three different inspection programs, IMC-2502, -2503 and -2504, during the same time frame. The inspection staff needed for the planning of the inspections should start to be assembled just before, or at least no later than, the docketing of the application and initiation of the licensing review.

The inspection planning function could operate from either the cognizant regional office or NRC headquarters, and should be conducted by the inspectors and scheduler selected to staff the resident inspectors' office at the site. This activity would be expected to shift to the site with the advent of significant safety-related site construction activities needing inspection coverage.

As stated above, construction inspection activities will be performed off site, as well as on site, and could involve inspections from at least three separate inspection programs. Therefore, there could be two or more separate organizations involved with inspection implementation for a single project. Remote inspections/audits of component and module fabrication activities, as well as design and engineering activities, could be conducted by different implementing organizations that would coordinate with the onsite construction inspection team using the CIPIMS. Examples of some of these inspections are as follows:

- First-of-a-kind engineering inspections under IMC-2502

- Inspections of detailed design information provided in place of DAC for approved designs (e.g., instrumentation and human factors of control room design), under IMC-2502 (Note: The process for resolving DAC prior to issuance of a COL is being considered as part of the 10 CFR Part 52 update rulemaking)

- ITAAC inspections under IMC-2503 for site construction activities

- ITAAC inspections under IMC-2503 for remotely manufactured modules or components

- Inspection of programs under IMC-2504

- Special Inspection programs

Special inspections/audits and foreign manufacturer inspections/audits are expected to be performed, or at least coordinated, out of headquarters.

An inspection program like the vendor inspection program would likely be needed for remote inspection/audit activities. These activities would be coordinated by either the Director of the Division of Inspection Program Management in NRR, by the lead region Division of Projects, or by a regional "center of excellence."

An inspection program, similar to IMC-2530, Integrated Design Inspection Program, should be developed for the inspection of first-of-a-kind (FOAK) engineering for the lead plant. A management decision would be needed to determine whether this FOAK engineering inspection would be conducted, or led, by the primary Region for the lead plant, the Region in the geographic location of the designer, a Region with a center of excellence in the area of design, or an NRC inspection group located in headquarters.

The October 1996 "Report on the Revised Construction Inspection Program" provided valuable insights into how an appropriate organization may look. In addition, the CIP team has developed the proposed organization described in Appendix G, "Example of an NRC Organization for Implementing a CIP," which parallels that of the Special Projects organizations of the late 1980s and early 1990s for TVA and Comanche Peak. Upon notification of the pending submittal of a COL application, these examples as well as the issues discussed above should be reviewed and evaluated in the context of a contemporary NRC organization to ensure that the CIP will be effectively and efficiently implemented.

3.6.3.4 ITAAC Comparison to Inspection Procedures

As part of an effort to determine how much of the old construction inspection program can be applied to future plants, the staff has reviewed the ITAAC that were developed for the ABWR and the AP600. A similar effort was previously done for the ABWR in SECY-94-294 for the high-pressure core flooder system. The staff broadened this effort to include all of the ITAAC for both plants to identify what inspection procedures will be used for an ITAAC and begin to estimate the amount of work involved in writing and rewriting the inspection procedures. Appendix H to this framework document contains the results of this review.

3.6.6.5 Inspection Findings and Enforcement

All inspection findings identified during the new construction period will be documented in accordance with IMC-0613 after they have been placed in context and assessed for impact on ITAAC. For enforcement purposes, the new construction period starts once the COL is approved for the facility and ends when the unit enters power operations. Once a COL is applied for, some regulations apply and enforcement actions may be considered for identified violations. During this period, potential violations from inspection activities will be processed in accordance with 10 CFR Part 2, the Enforcement Policy, and other applicable enforcement guidance using traditional enforcement tools. The findings will then be categorized as violations, deviations, non-conformances, or unresolved items. This includes use of severity levels, notice of violations (NOVs) for violations of severity level III and above, and civil penalties as appropriate.

Once the facility enters power operations, there will be a transition to the reactor oversight process (ROP). During this transition period, inspection findings and enforcement actions will be processed using the ROP as much as practicable. The approach for transitioning to the ROP will be as follows:

The facility transition to the ROP will be a gradual-phased approach on an individualized cornerstone basis. The basis for determining that a cornerstone is ready to be monitored under the ROP will be documented. The document will contain all the records that verify that a cornerstone can be monitored fully.

When one or several cornerstones appear to be ready to be monitored under the ROP, a transition plan will be developed which will specify which inspections will be performed to verify that all issues have been resolved and that all licensee corrective actions are effective.

The regulatory response and plant performance assessment will be in accordance with the Action Matrix as defined in IMC-0305, "Operating Reactor Assessment Program." During the transition to the ROP, the regulatory responses allowed by the Action Matrix may be used with the concurrence of the management team assigned responsibility for construction and the regional administrator.

The transfer of the facility to the full reactor oversight process will be accomplished by written approval of the regional administrator with the concurrence of the Director, Office of Nuclear Reactor Regulation (NRR). This transfer may occur even if all performance indicators are not yet available, provided compensatory inspections are conducted as provided for by IMC-2515. The management team assigned responsibility for construction may be dissolved at that time or may be maintained for up to two additional quarters if necessary to deal appropriately with outstanding issues.

4. PUBLIC INVOLVEMENT

Public involvement has been an important aspect of the development of this document. Every effort has been made to seek input from both internal and external stakeholders during the process of developing the scope and defining the content of the construction inspection program for plants that might be built under 10 CFR Part 52. This document was reviewed by the Advisory Committee on Reactor Safeguards (ACRS) in December 2003. The recommendations made by the ACRS (ADAMS Accession Number ML033460266) have been incorporated.

Public comment on this document was sought through a *Federal Register* Notice in May 2003, when the framework was initially proposed. A subsequent workshop in August 2003, offered the opportunity for the NRC staff to provide more in-depth descriptions and explanations of the activities planned under the CIP and detailed in the framework. The workshop also provided an opportunity for external stakeholders to ask questions and suggest alternatives for various aspects of the framework. Ideas and issues raised during the workshop were captured in a written transcript of the day-long meeting.

Written comments submitted in response to the *Federal Register* Notice as well as suggestions and questions raised during the workshop were considered and incorporated to the extent possible into this version of the framework. A detailed listing of the various comments as well as an explanation of how the comment was resolved can be found in Appendix I, "Comment Resolution Summary for the Draft Construction Inspection Program Framework Document."

Appendix A
Glossary

<u>Attribute Guidance</u>. The Inspection Guidance, generally discussed in each NRC inspection procedure, that relates to the types of activities an inspector should observe and review and together with some references, providing specific acceptance criteria that can be used in the evaluation process.

<u>Audit</u>. An applicant/contractor activity to determine through investigation the adequacy of, and adherence to, established procedures, instructions, specifications, codes, and other applicable contractual and licensing requirements, and the effectiveness of implementation

<u>Construction Inspection Program Information Management System (CIPIMS)</u>. CIPIMS is a dedicated, computer-based inspection scheduling and information management system intended for deployment at nuclear power plants (NPPs) under construction. CIPIMS will be used to integrated the inspection schedule with the licensee's construction schedule. It will also be used to organize and manage information about the inspection results, the licensee's ITAAC completion information, and the NRC's ITAAC determinations.

<u>Combined License</u>. A combined construction permit and operating license with conditions issued pursuant to 10 CFR Part 52. Like a construction permit under 10 CFR Part 50, a combined license under 10 CFR Part 52 authorizes construction of a nuclear power plant. The NRC ensures that the licensee has completed the required inspections, tests, and analyses, and authorizes operation after finding that the acceptance criteria have been met.

<u>Construction Activities</u>. Any activity associated with the construction, fabrication, or testing of structures, components, subcomponents, subsystems, or systems either at the construction site or at remote fabrication or testing facilities that occurs during the construction phase of the inspection program. Construction activities also include the design and engineering of the structures, systems, and components of the facility.

<u>Contractor</u>. Any organization under contract for furnishing items or services to an organization operating under the requirements of Appendix B of 10 CFR 50 or the commitments made in the application. It includes the terms Consultant, Vendor, Supplier, Fabricator, Constructor, and subtier levels of these, where appropriate.

<u>Critical Attribute</u>. A characteristic or quality of any construction material, object, action, or process that demonstrates that design and performance requirements have been met either uniquely for the item or collectively for the related structure, system, or component. Critical attributes (which may also be applicable to construction documents, such as procedures, reports, and records) are discussed in each NRC inspection procedure as Inspection Requirements, delineating specific inspection activities that may be conducted to check the listed attributes for conformance with the relevant acceptance criteria.

Design Control Document. The design control document is a repository of information on the respective standard plant design (e.g., AP600, advanced boiling water reactor). The design control document also provides the design-related information that is incorporated by reference into the respective appendix to 10 CFR Part 52. The design control document consists of Tier 1 and Tier 2 information (see below for definitions).

Documentation. Any written or pictorial information describing, defining, specifying, reporting, or certifying activities, requirements, procedures, or results.

Early Site Permit. Under 10 CFR Part 52, an early site permit addresses site suitability issues, environmental protection issues, and plans for coping with emergencies, independent of the review of a specific nuclear plant design.

Inspection. (1) An NRC activity consisting of examination, observation or measurements to determine applicant/contractor conformance with requirements and/or standards. (2) Applicant/contractor quality control measures consisting of examination, observation or measurements to determine the conformance of materials, supplies, components, parts, appurtenances, systems, processes or structures to pre-determined quality requirements.

Inspection Finding. A documented evaluation of the acceptability of licensee construction activities.

Inspection Sample. An item selected for inspection of one or more critical attributes. For example, an inspection sample may be a single record for review of welding records, while an entire system would comprise the inspection sample during a system walkdown inspection. The composition of an inspection sample will be defined in each inspection procedure under the sampling criteria. The inspection sample should be identified in CIPIMS with the licensee's unique identification number.

Inspections, Tests, Analyses, and Acceptance Criteria (ITAAC). ITAAC are a provision of the Atomic Energy Act and 10 CFR Part 52. They are necessary to ensure that a plant licensed in accordance with 10 CFR Part 52 has been properly constructed and will operate safely. The licensee performs the ITAAC and the Commission must find that the ITAAC have been met before fuel loading at the nuclear power plant is allowed.

ITAAC Determination. A determination about the completion of an ITAAC that is made by the inspection staff after reviewing the inspection history and the licensee's documentation related to the ITAAC. This determination is performed for individual ITAAC and when combined for all ITAAC will lead to a recommendation that the Commission makes a finding in accordance with 10 CFR 52.103(g).

Lead Region. The region designated with the authority to make a recommendation to the Director of Nuclear Reactor Regulation that an activity has been satisfactorily completed based on inspections associated with an early site permit or combined license application. The lead region is based on geography and is defined as that region that oversees the location of an early site permit or combined license.

Limited Work Authorization (LWA). Authorization from the NRC to an applicant to conduct certain construction activities pursuant to 10 CFR 50.10(e)(1) or 10 CFR 50.10(e)(3)(i).

Sign As You Go (SAYGO) Process. For selected construction activities, the NRC will perform inspections beginning during the early stages of reactor construction to assure that construction activities are accomplished in accordance with licensee procedures, applicable codes and standards, and NRC regulations. In addition, the NRC will check to ensure that the licensee has implemented QA/QC oversight of these activities such that acceptable quality is consistently maintained. If the activities are resulting in consistently satisfactory results, the NRC will 'sign-off' on the activity and will consider reducing the inspection effort in that area.

Tier 1 Information. Tier 1 information is that portion of the design-related information in the design control document that is approved and certified by the NRC through rulemaking. Tier 1 information includes the following:
- definitions and general provisions
- design descriptions
- ITAAC
- significant site parameters
- significant interface requirements

Tier 2 Information. Tier 2 information is that portion of the design-related information in the design control document that is approved but not certified by the design certification rule. Tier 2 information includes the following:
- information required by 10 CFR 52.47, with the exception of generic technical specifications and conceptual design information
- information required for a final safety analysis report under 10 CFR 50.34
- supporting information on the inspections, tests, and analyses that will be performed to demonstrate that the acceptance criteria in the ITAAC have been met
- combined license information items which identify certain matters that shall be addressed in the site-specific portion of the final safety analysis report by an applicant who references a design certification rule

Appendix B
Information Considered in Updating
the Construction Inspection Program

This appendix discusses lessons learned from the 1996 revised construction inspection program document which was considered in the development of the 10 CFR Part 52 Construction Inspection Program (CIP). The report covers a variety of programs, activities, and experiences from the last NRC construction inspections conducted at Seabrook, Comanche Peak, South Texas, Watts Bar, and Bellefonte. In updating the CIP, the staff considered the previously reviewed foreign construction inspection practices and the modular construction techniques used in the US shipbuilding industry.

Quality Processes

- The assessment process must begin with inspections of the design engineering process, including engineering quality assurance (QA), to ensure that the licensee can accurately translate high-level design requirements into detailed engineering and fabrication drawings.

- Because NRC inspections are done on a sampling basis, the CIP must provide accurate assessment of the licensee's quality programs. To the extent possible, all construction inspections should assess the effectiveness of QA and quality control (QC), and the results must be thoroughly documented and integrated. Additionally, the staff intends to perform programmatic QA inspections to provide reasonable assurance that Appendix B requirements are adequately implemented. Ideally, the breadth and depth of the NRC's verification that a plant's QA and QC are effective will ensure that any demonstrated or alleged lapses in quality are isolated instances rather than generic problems.

- The licensee's management of QC records is an integral part of the quality process. In order to verify the overall adequacy of the licensee's QA records management process, the CIP must inspect all aspects of QA/QC records, from creation through storage.

- The identification of construction problems and the timeliness and extent to which they are corrected are effective measures of licensee management's control over onsite activities. NRC experience shows that, if the licensee has a thorough corrective action program and effectively identifies and corrects root causes of problems, there is a good chance that the overall quality of the construction is good. If the corrective action program is weak, it is likely that there are lapses in quality (i.e., if repetitive problems occur).

Inspection Program Management

The objectives of the inspection program is to support the Commission's 52.103(g) finding that all ITAAC have been met and that programs are in place to ensure the facility will operate in conformance with the Commission's regulations. This approach will be more likely to produce enough inspection data to assess the adequacy of a plant's construction and readiness to commence operations. These objectives should be considered in establishing the inspection methodologies to be employed (e.g., risk-informing the inspection sample selection, inspection sample size) and the format and content of inspection documentation.

- In the past, NRC construction inspections were often scheduled on the basis of inspector availability. Inspections were therefore performed on activities that happened to be in progress at the time of the inspection, resulting in a less-than-optimum sample selection. Because inspectors will be continuously onsite under the revised CIP, and because ITAAC must be verified under Part 52, NRC inspections must be scheduled on the basis of the utility's construction progress. All aspects of the construction inspection program, including inspection planning, scheduling, preparations, and implementation, must be conducted in a way that will ensure all necessary attributes are properly inspected.

- The proper mix of skills and experience among inspectors, particularly during the near-term operating license (NTOL) phase at a plant, is necessary to ensure effective implementation of the inspection program.

- The CIP must be able to support NRC action on a licensee's certification of readiness to load fuel, all ITAAC having been completed satisfactorily. The inspection staff should be fully aware, in advance, of all issues the licensee will address in its certification.

- Inspection results must be assessed to verify that inspection requirements are met and that the results support the objectives of individual inspection procedures and of the CIP.

- A plan for the transition from the construction phase to the operations phase should be made well in advance of the completion of plant construction. This transition plan, which can be viewed as an exit strategy for exiting the CIP, should be based on projected inspection workload and must provide for the necessary turnover of issues.

- It is necessary to ensure that each phase of the preoperational inspection program is properly completed. To the maximum extent possible, all issues (such as licensee test exceptions or construction deficiencies) must be closed out before the programs are officially considered complete. Items that are carried over into the operating phase must be extensively documented, and the closure requirements for the items must be identified.

Inspection Program Structure and Implementation

The program must be structured to guide inspectors to inspect needed items and to provide a coherent and simple method for them to record necessary information.

- Onsite inspections should begin during site preparation before the COL or CP is issued. A continuous onsite inspection staff should be established and maintained throughout construction. To ensure that the full range of construction activities is covered by appropriately qualified inspectors, and because of the phased nature of many of those activities, the mix of expertise among the resident inspection staff should be rotated.

- Inspection requirements should be made as objective as possible, allowing clear determinations that critical attributes either have or have not been met. Establishing

B-2

discrete, objective inspection requirements will limit the need for subjective interpretations of acceptability, and the sizable body of accumulated objective information will support major inspection program conclusions.

- Objective inspection requirements should be established, to the maximum possible extent, for systems, structures, and components, as well as for plant programs. Each inspection procedure should clearly state how much inspection should be performed in order to consider the procedure complete.

- Constructing a plant in a short period of time means that activities will happen rapidly and in parallel with each other, which will place significant demands on inspection resources. Planning and scheduling therefore need to be closely coordinated with plant construction plans.

Inspection Documentation

At the end of the construction process, NRC must possess a fully documented body of inspection data to support the findings that need to be made to allow plant operation.

- In some past construction projects, inspection reports did not fully document all areas that had been evaluated during plant construction. The resulting incomplete inspection documentation resulted in a lack of audit trails that could be used to respond to questions raised during the process leading up to issuance of an operating license. Also, inspection reports did not always clearly identify the items that had been inspected in the plant. The revised CIP requires those individual samples (such as identification numbers for welds, pipe supports, and cable terminations) be recorded in the CIPIMS. In addition, each construction inspection in the future should be considered satisfactorily completed only after supervisory or management personnel determine that the inspection is fully documented.

- In the past, NRC inspection reports focused generally on the deficiencies identified during the inspections, without providing much detail on positive inspection findings. As a result of such unbalanced inspection reporting, the NRC staff sometimes had to perform extensive reviews during the final stages of plant licensing to provide additional information to support licensing decisions. In some cases, the inspections had already been done but had not been fully documented. To avoid follow-up reviews, future construction inspections should document both satisfactory and unsatisfactory findings.

Appendix C
Inspection Sampling

<u>ITAAC Sample Selection</u>

A cornerstone of the Part 52 process is the concept of ITAAC which if met are necessary and sufficient to provide reasonable assurance that the facility has been constructed and will be operated in conformity with the license, the provisions of the Atomic Energy Act, and the Commission's rules and regulations. The staff will rely on the licensee to ensure that all of the ITAAC have been met and will perform audit-type inspections to verify compliance with the ITAAC. In performing these audit-type inspections the staff needs to address the fundamental question of how much inspection is necessary to ensure that the acceptance criteria contained in the ITAAC have been met.

Both the licensee and the NRC benefit from the process used during the COL and design certification reviews that determine the level of detail for the ITAAC. For the designs that have been certified (i.e., ABWR, System 80+, and the AP600) ITAAC were developed for the SSCs within the scope of the designs. These ITAAC are part of the Tier 1 material found in the design control document for these designs.

A process was developed to determine the level of detail for each ITAAC. This process is discussed in sections 14.3 of the respective design control documents and also in the staff's draft standard review plan section 14.3. The Tier 1 information has an entry for every system that is either fully or partially within the scope the design certification. The intent of this comprehensive listing is to define at the Tier 1 level the full scope of the certified design. However, the amount of information in the Tier 1 entry, including the ITAAC, is commensurate with the significance of the system. Several factors were used to determine the significance of the system including the following:

- whether the feature or function is necessary to satisfy the NRC's regulations in 10 CFR Parts 20, 50, 52, 73, or 100
- whether the feature or function pertains to a safety-related structure, system or component
- whether the feature or function represents an important assumption or insight from the probabilistic risk assessment
- whether the feature or function is important in preventing or mitigating a severe accident
- whether the feature or function has had a significant impact on the safety or operation of existing nuclear power plants
- whether the feature or function is typically the subject of a provision in the technical specifications
- whether the feature or function in question is specified in the standard review plan as being necessary to perform a safety-significant function

For many non-safety systems with low risk significance, the Tier 1 entry is limited to the systems' name only. For this group, it is sufficient to ensure that the system has been completed before fuel loading is allowed.

The staff believes that the process that has been developed and implemented for the certified designs provides a good starting point for answering the fundamental question of how much inspection is enough. That is, by having a construction inspection program that is ITAAC-focused for the hardware portion of the design, the staff has already narrowed the field of inspection activities. Operational programs, which do not have ITAAC associated with them are another matter and are discussed in other sections of this document.

The staff does not intend to review or inspect every inspection, test, or analysis listed in every ITAAC. To establish an NRC inspection footprint, the staff will ensure that, at a minimum, it has received an ITAAC determination letter from the licensee for all ITAAC. If no inspections have been performed related to that ITAAC as documented in CIPIMS (e.g., inspection findings or assessment) the staff will review the licensee's records for the ITAAC determination basis as necessary to provide confidence that the ITAAC have been met.

Statistical Methods

This approach involves the development and implementation of statistical sampling methods with the goal of obtaining, at the end of a plant's construction phase, a confidence statement about the quality of plant construction. The October 1996, draft revised CIP report noted that the major difficulty with applying statistical sampling to a nuclear power plant construction inspection program would arise from the attempt to make confidence statements about the many non-homogeneous processes that occur at a construction site.

The draft revised CIP report also referenced a memorandum to the Commission from E. Volgenau, Director, Office of Inspection and Enforcement dated February 11, 1977, titled, "Inspection Program Utilizing Statistical Sampling Inspection Techniques." This memorandum discussed the results of a series of statistically based operating phase inspections that were performed at Three Mile Island Unit 1. This trial program showed that strictly statistically based sampling was, on balance, not an optimal method of inspection planning for three reasons:

- the statistical method identified no significant safety concerns that the traditional method failed to identify;
- the traditional method successfully identified significant safety concerns that the statistical method did not identify,
- and; the statistically based method was comparatively more resource-intensive.

However, the memorandum did note that confidence statements for a wide range of populations and sample sizes could be developed for possible application to discrete portions of the inspection program.

Since the time of the 1977 memorandum the staff has applied the use of statistical sampling techniques to inspection-related activities. For example, resolution of some issues associated with construction inspection of welding programs, and the dedication of commercial-grade items for use in nuclear power plants relied on the use of statistical sampling techniques.

Regarding welding programs, shortly before the Seabrook full-power license was to be issued, the NRC received a series of allegations, questions, and concerns about safety at the plant. Some of the issues related to the adequacy of pipe welds made on-site during construction. The NRC used statistical based sampling techniques to aid in its investigations and inspections of this issue. An October 4, 1991, letter from James Taylor to the Commission titled, "Completion of the NRC Staff Review of the Quality of ASME Field Welds at Seabrook," references these techniques.

Sampling techniques are also discussed in Draft Regulatory Guide 1070, "Sampling Plans Used for Dedicating Simple Metallic Commercial Grade Items for Use in Nuclear Power Plants." The commercial-grade dedication in this draft regulatory guide refers to an acceptance process undertaken to provide reasonable assurance that a commercial-grade item to be used as a basic component in a nuclear power plant will perform its intended safety function and is deemed equivalent to an item designed and manufactured under a quality assurance program in accordance with 10 CFR 50 Appendix B. Although the draft regulatory guide is intended to provide guidance for the development of a licensee's commercial-grade dedication programs, the staff believes that some of the concepts developed for this program are applicable to the NRC's own construction inspection verification programs.

An area where statistical sampling techniques could possibly be used for the construction inspection program is welding. Both the ABWR and the AP600 contain ITAAC associated with welding. Statistical-based sampling techniques could be used for the staff to make findings and ITAAC determinations for this process. If future plants were to have welding performed off-site (because of modular construction techniques) and on-site, separate welding assessments could be made for the off-site facilities as well as for on-site welding activities.

The staff could use statistical sampling techniques such that it will have a high confidence of a low defect rate. To satisfy this premise, both the resolution of issues associated with the Seabrook welding issue and DG-1070, used statistical sampling plans that result in at least 95 percent confidence that populations with more than 5 percent defective items will be rejected. Because of modular construction techniques it may be necessary to make assessments about several facilities taken together. If this is the case then the samples at each facility can be adjusted to support the staff's evaluation.

The staff believes that such statistically based sampling techniques are limited to certain areas. As the revision to the construction inspection program moves forward, the staff hopes to identify ITAAC and construction activities that lend themselves to such a technique and develop inspection procedures that will provide guidance for how such techniques should be used.

Risk Informing Construction Inspection

The 1996 report on the revised CIP identified that PRA information could be used by the NRC to perform sensitivity, uncertainty, and importance analyses to identify those plant SSCs whose passive failure (due to inadequate construction) would most greatly impact the plant's risk profile. In this way, the more risk-significant SSCs would be identified, and construction inspection samples could be skewed toward those SSCs.

The selection of ITAAC were heavily risk informed during the design certification process because design-specific PRA is required as part of a design certification in accordance with 10 CFR

C-3

52.47(a)(v). These PRAs were used during the applicant's development, and the staff's review, of the ITAAC for the designs that were certified.

The Use of Risk in Developing ITAAC for AP600

The AP600 was chosen as an example because the safety systems for this design use passive means (such as gravity, natural circulation, condensation and evaporation, and stored energy) for accident prevention and mitigation. These passive safety systems perform safety injection, residual heat removal and containment cooling functions. In this design, traditional active systems like the emergency diesel generators are non-safety related.

Section 14.3 of the AP600 design control document provides background information on the selection criteria for how the ITAAC were developed. The selection criteria consisted of deterministic and PRA based inputs. Table 14.3-1 of the AP600 design control document provides the results of the ITAAC screening summary. The screening of the 90 AP600 systems led to several systems not being selected for an ITAAC. In addition, for the AP600 there are 32 systems, such as the potable water system, turbine building closed cooling water system, and the heater drain system, were only the system is listed in the Tier 1 material. The end result is that 39 of the original 90 systems (greater than 40%) that were screened resulted in no detail ITAAC being developed.

While many systems were screened out for consideration during the ITAAC development several systems were included and different aspects of those systems augmented in the ITAAC because of risk insights. As mentioned earlier the emergency diesel generator is non-safety related for the AP600 design, however, there are ITAAC associated with the EDG because of it's risk significance. Similarly, there are non-safety related functions of the normal residual heat removal system that have ITAAC associated with them, in part because of their risk significance. Table 14.3-6 of the AP600 design control document contains the design features from the PRA perspective that were considered important to verify in ITAAC. Because of the information in the design certification the staff has a good starting point for the use of risk in the inspection program for the designs that have been certified.

The staff intends to use the design-specific PRA's to help further focus its ITAAC inspection activities. While such risk information will be useful in developing construction inspection samples and focusing on audit activities, the actual conduct of construction inspections will primarily represent a deterministic process. This is important because a plant must be built in accordance with its design criteria for subsequent PRA usage to be valid.

The Role of the Quality Assurance Program

Quality assurance (QA) and quality control (QC) will be an integral part of the NRC's inspection effort, and will be a common component of the inspections that are performed by the staff, in that 10 CFR 50 Appendix B applies to construction activities done in accordance with 10 CFR Part 52. The staff believes that one of the major lessons learned from past nuclear power plant construction efforts is that the identification of construction problems, and the timeliness and extent to which they are corrected are effective measures of licensee management's control over onsite activities.

NRC experience shows that, if the licensee deals thoroughly with corrective action, including the identification and correction of root causes, there is a good chance that the overall quality of the construction is good. If these areas are weak, it is likely that there are lapses in quality; such a case would be evident if repetitive problems occur.

The role of quality assurance was emphasized in SECY-00-0092, "Combined License Review Process," dated April 20, 2000. The following is paraphrased from this SECY paper:

> The NRC staff anticipates that there will be design, construction, and testing activities related to ITAAC verification for which the staff will not be able to rely solely on NRC inspections to verify proper completion. For these activities, the staff must rely on the licensee's QA program to provide suitable controls for effective verification. The staff must have confidence that the licensee's QA program is adequate and that it is being properly implemented so that design, construction, or testing deficiencies are identified, documented, and corrected. The QA requirements of Appendix B to Part 50 apply to all safety-related activities being conducted by the licensee during the design, construction, and operations phase, including those safety-related activities performed to satisfy ITAAC. For example, preoperational test program testing performed to demonstrate that safety-related structures, systems, and components (SSCs) will perform satisfactorily in service must be conducted under a program that satisfies Criterion XI, "Test Control," of Appendix B. It may also satisfy testing required by the ITAAC process. The scope of the initial test program, however, is not limited to just safety-related SSCs. Specifically, Regulatory Guide (RG) 1.68, "Initial Test Programs for Water-Cooled Nuclear Power Plants," specifies the scope of plant SSCs to be tested to satisfy the requirements of Criterion 1, "Quality standards and records," of Appendix A , and Appendix B to Part 50. Although testing is required for all SSCs within the scope of RG 1.68, it is not required that all of them be tested to the same stringent requirements. Accordingly, the administrative requirements that govern the conduct of the test program contain provisions for the application of administrative controls in a manner commensurate with the safety significance of the SSCs within its scope. Because the ITAAC process includes safety-related activities that must be conducted under a QA program that meets the requirements of Appendix B to Part 50, licensees must develop programmatic controls and procedures that delineate how such activities will be implemented.

> As discussed in public meetings with NEI representatives, there may be deficiencies identified by the QA program that are relevant to ITAAC and that must be addressed by the licensee before the NRC can find that the ITAAC have been successfully completed. NEI representatives asserted that QA and QC deficiencies have no relevance to ITAAC findings. The NRC staff disagrees with any assertion that QA/QC deficiencies have no relevance to the determination of whether ITAAC have been successfully completed. Simply confirming that ITAAC had been performed in some manner and a result obtained apparently showing that the acceptance criteria had been met would not be sufficient to support a determination that ITAAC had been successfully completed. The manner in which ITAAC are performed can be relevant and material to the results of the ITAAC. For example, in conducting

ITAAC to verify a safety-related pump's flow rate, it is necessary, even if not explicitly specified in the ITAAC, that the gauge or instrument used to verify the pump flow rate be calibrated in accordance with the requirements of Appendix B to Part 50 and that the test configuration be representative of the final as-built plant conditions (i.e., valve or system lineups, gauge locations, system pressures, or temperatures). Otherwise, the acceptance criteria for pump flow rate could apparently be met while the actual flow rate in the system could be different than that required by the approved design. Therefore, the NRC staff has determined that a QA/QC deficiency may be considered in determining whether an ITAAC has been successfully completed if (1) the QA/QC deficiency is directly and materially related to one or more aspects of the relevant ITAAC (or supporting Tier 2 information) and (2) the deficiency (considered by itself, with other deficiencies, or with other information known to the NRC) leads the NRC to question whether there is a reasonable basis for concluding that the relevant aspect of the ITAAC has been successfully completed. This approach is consistent with the NRC's current methods for verifying initial test programs.

The NRC staff recognizes that there may be programmatic QA/QC deficiencies that are not relevant to one or more aspects of a given ITAAC under review and, therefore, should not be relevant to or considered in the NRC's determination as to whether that ITAAC has been successfully completed. Similarly, individual QA/QC deficiencies unrelated to an aspect of the ITAAC in question would not form the basis for an NRC determination that an ITAAC has not been met. Using the ITAAC for pump flow rate example, a specific QA deficiency in the calibration of pump gauges would not preclude an NRC determination of successful ITAAC completion if the licensee could demonstrate that the original deficiency was properly corrected (e.g., analysis, scope of effect, root cause determination, and corrective actions, as appropriate) or that the deficiency could not have materially affected the test in question. Furthermore, during the development of ITAAC, the design certification applicants determined that it was impossible (or extremely burdensome) to provide all details relevant to verifying all aspects of ITAAC (e.g., QA/QC) in Tier 1 or Tier 2. Therefore, the NRC staff accepted the applicants' proposal that top-level design information be stated in the ITAAC to ensure that it was verified, with an emphasis on verification of the design and construction details in the "as-built" facility. To argue that consideration of underlying information, which is relevant and material to determining whether ITAAC have been successfully completed, is not necessary ignores this history of ITAAC development.

In the September 5, 2002, staff requirements memorandum associated with SECY-00-0092, the Commission approved the staff's recommendation that underlying information (such as QA/QC deficiencies), which is relevant and material to ITAAC, must be considered in determining whether ITAAC have been successfully completed. In addition, there may also be deficiencies identified that are not relevant to ITAAC. These deficiencies may still need to be addressed by the licensee, but they will not necessarily delay a finding on successful ITAAC completion or plant operation.

In summary, the staff believes that statistical sampling and PRA techniques can be used as an aid to help to focus its ITAAC-based inspection efforts. The staff also believes that inspections of

QA/QC (especially corrective action program inspections) will be an important aspect of the review. This paper presents a high-level approach for using these methods. Details of the design need to be known to employ these techniques properly.

The staff plans to take selected examples from the designs that have been certified and develop these techniques further. The results of these examples will be made publicly available and will be used as aids in the development of the detailed inspection procedures for a COL. The staff intends to delay work on revising the detailed inspection procedures until it has more information from the industry on the details and the design for any particular nuclear power plant that may be constructed in the future.

Appendix D
Examples of Information That Would
Invalidate a Previous ITAAC Determination

This appendix gives examples for discussion of what the NRC staff considers what constitutes"new and significant information." The examples show how "new and significant information" might impact a previously accepted ITAAC.

Example 1: A test instrumentation QA/QC deficiency directly related to whether an ITAAC acceptance criterion had been met.

The role of quality assurance was emphasized in SECY-00-0092, "Combined License Review Process," dated April 20, 2000. The following is paraphrased from this SECY paper.

The manner in which ITAAC is performed can be relevant to its results. For example, in verifying an ITAAC associated with a safety-related pump's flow rate, the gauges or instruments used to verify the pump flow rate must be calibrated in accordance with the requirements of 10 CFR Part 50 Appendix B. The test configuration must also be representative of the final as-built plant conditions. For example, valve or system lineups, gauge locations, system pressures, and temperatures must be in accordance with the design. Otherwise, the acceptance criterion for a pump flow rate could apparently be met while the actual flow rate in the system was different from the required design flow rate. The NRC staff therefore determined that a QA/QC deficiency may be considered in determining whether an ITAAC has been successfully completed.

A QA/QC deficiency could be relevant if the deficiency is directly related to the ITAAC or its supporting Tier 2 information. The deficiency by itself or with other deficiencies, may lead the NRC to question whether there is a reasonable basis for concluding that the relevant aspect of the ITAAC has been successfully completed. This approach is consistent with the NRC's current methods for verifying initial test programs.

Example 2: Improper weld materials used in the fabrication of an ITAAC related structure. (from Appendix E)

The example involves the reactor pressure vessel support platform for Taipower's Lungmen-1 ABWR. Improper welding material was used for initially assembling the platform. The 1,000-ton platform support was made of steel-reinforced concrete and the steel portion was manufactured at the China Shipbuilding Corp. (CSC) in Kaohsiung in southern Taiwan. The platform was shipped to the site and then assembled by CSC personnel at the Lungmen site.

Workers initially used low-strength welding material to assemble the platform, instead of high-strength material specified by the engineering codes. The welding material was confirmed to be inappropriate and Taipower will have to reassemble the platform.

The information concerning the use of improper welding material could have come in the form of an allegation to the licensee, or to the NRC. In such a case, the NRC would investigate and if the allegation were substantiated the NRC would evaluate the impact on any associated ITAAC.

Appendix E of this document discusses this example in more detail including the ITAAC that may be impacted.

Example 3: IE Bulletin No. 83-07, Apparently Fraudulent Products Sold by Ray Miller, Inc., and NRC Bulletin No. 88-05, "Nonconforming Materials Supplied by Piping Supplies, Inc. (PSI) at Folsom, New Jersey and West Jersey Manufacturing Company (WJM) at Williamstown, New Jersey."

Bulletin No. 83-07 was issued after the NRC completed a review of records that were in the custody of the U.S. Attorney's office, and determined that materials with fraudulent documentation had been supplied to nuclear power plants.

Bulletin No. 88-05 was issued after the NRC obtained copies of certified material test reports (CMTRs) for material supplied by PSI and WJM that contain false information about material supplied to the nuclear industry. A domestic forging company's letterhead was apparently used on a number of CMTRs to certify that commercial-grade and foreign steel met the requirements of ASME Code Section III, Subarticle NCA-3800. There was no evidence that PSI or WJM performed or had a subcontractor perform the testing required by Section III to upgrade the commercially produced steel for these falsified CMTRs.

Information of this nature concerning the construction materials for ITAAC-related components or structures could be considered significant enough to invalidate a previous ITAAC determination.

Example 4: NRC Bulletin No. 92-01, "Failure of Thermo-lag 330 Fire Barrier System to Maintain Cabling in Wide Cable Trays and Small Conduits Free From Fire Damage"

This bulletin notified licensees of failures in fire endurance testing associated with the Thermo-Lag 330 fire barrier system installed to protect safe shutdown capability, and requested all operating reactor licensees to take recommended actions.

During construction, information of this nature concerning fire protection materials for safety-related or risk-significant systems could be considered significant enough to invalidate a previous ITAAC determination.

Example 5: NRC Bulletin No. 88-10, "Nonconforming Molded-Case Circuit Breakers"

NRC Information Notice (IN) 88-46, "Licensee Report of Defective Refurbished Circuit Breakers," and Supplement 1 thereto, reported that Anti-Theft Systems, Inc., a local electrical distributor, supplied 30 circuit breakers (CBs) to the Diablo Canyon nuclear power plant. These circuit breakers (Square D molded-case, type KHL 36125) were intended for use in non-safety-related applications at Diablo Canyon. Square D Company reported that inspection and testing of these CBs determined that they were refurbished Square D Company equipment. Furthermore, Square D reported that several of the circuit breakers tested did not comply with Square D or Underwriters Laboratories, Inc. (UL) specifications for all of the electrical tests performed. IN 88-46 also listed several California companies that were involved in supplying surplus, and possibly defective, refurbished electrical equipment to the nuclear industry.

During the NRC inspections of defective refurbished circuit breakers, additional examples were identified that indicate a potential safety concern regarding electrical equipment supplied to nuclear power plants. The NRC was concerned that equipment being procured as new, and assumed to meet all applicable plant design requirements and/or original manufacturer's specifications may not conform to these requirements and specifications.

While the bulletin discussed CBs supplied for non-safety-related applications, it is now understood that non-safety-related electrical systems can be risk-significant. Information of this nature, concerning potentially defective circuit breakers for safety-related or risk-significant systems, could be considered significant enough to invalidate a previous ITAAC determination.

Appendix E
ABWR Construction Example

This appendix provides an example of a construction activity that could be done off site that could impact inspections, tests, analyses and acceptance criteria (ITAAC). The example involves the Lungmen-1 advanced boiling water (ABWR) reactor that is being constructed in Taiwan. The design is very similar to the design that was certified by the NRC and that is codified in Appendix A of 10 CFR Part 52. However, because of licensing differences, there are no ITAAC associated with the Lungmen design. The staff believes that if the problem that was encountered in Taiwan happened in the United States, it would directly impact an ITAAC.

Description of the Problem

The problem involves the reactor pressure vessel support platform for Taipower's Lungmen-1 ABWR. Improper welding material was used for initially assembling the platform. The 1,000-ton platform holding the vessel is made of steel-reinforced concrete and was manufactured at the China Shipbuilding Corp. (CSC) in Kaohsiung in southern Taiwan. The platform was shipped to the site and then assembled by CSC personnel at the Lungmen site.

Workers initially used low-strength welding material to assemble the platform instead of high-strength material specified by engineering codes. The welding material was confirmed to be inappropriate and Taipower will have to reassemble the platform.

The vessel support platform has five layers of reinforced concrete. The initial problem manifested itself when a hairline crack about 50 centimeters (cm) long and between 0.2 cm and 0.3 cm deep was discovered in the lowest level of the platform. The platform weighs 464 metric tons, is 13 meters (m) high and has a diameter of 14 m. At the time of the discovery Taipower filed a so-called quality assurance "noncompliance" report with the regulator. It was subsequently determined through an inspection that improper welding material was used for the assembly. The inspection was performed after irregularities in welding were suspected on the site.

U.S. Design Control Document Information

The design control document for the U.S. ABWR is incorporated by reference into Appendix A of 10 CFR Part 52. The design control document consists of Tier 1 and Tier 2 information and generic technical specifications. The Tier 1 material consists of the following:

- definitions and general provisions
- design descriptions
- ITAAC
- significant site parameters
- significant interface requirements

The following information is extracted from Section 2.14.1, "Primary Containment System," of the U.S. ABWR Tier 1 information:

> The RPV pedestal forms the lower drywell region and consists of a cylindrical double shell composite steel structure. It is anchored to the basemat and supports the RPV through a support ring girder. The pedestal also supports the reactor shield wall. The pedestal consists of two concentric steel cylinders joined together radially by vertical steel diaphragms and filled with concrete. The pressure suppression venting paths are an integral part of the pedestal structure, which includes (1) the ducts which interconnect the lower and upper drywell regions, (2) the vertical downcomers from the interconnecting ducts to the horizontal vents, and (3) the horizontal vents that direct steam into the suppression pool. The horizontal vents consist of 30 pipes uniformly spaced around the perimeter of the pedestal in ten stacks of three each. The total horizontal vent area is greater or equal to 11.55 m². The distance from the pedestal containing these horizontal vents to the outer suppression pool wall is greater than 7.4m. All HVAC ducts, cabling and piping between the upper and lower drywells are routed through the interconnecting ducts

The ITAAC that could be affected (if the problem occurred in the U.S.) are 2.14.1.3 and 2.14.1.14 (shown below).

Table 2.14.1 Primary Containment System

Design Commitment	Inspections, Tests, Analyses	Acceptance Criteria
3. The ASME Code pressure boundary components of the PCS will retain their integrity under internal pressures that will be experienced during service.	3. A structural integrity test (SIT) will be conducted on the pressure boundary components of the PCS per ASME Code requirements.	3. The results of the SIT of the pressure boundary components conform with the requirements of the ASME Code.
14. The containment internal structures are able to withstand the structural design basis loads as defined in Section 2.14.1.	14. A structural analysis will be performed which reconciles the as-built data with structural design as defined in Section 2.14.1.	14. A structural analysis report exists which concludes that the as-built internal structures are able to withstand the design basis loads as defined in Section 2.14.1.

A diagram of the ABWR primary containment system is also contained in the Tier 1 material as follows (note 2 in the figure refers to the location of the reactor pressure pedestal):

Lessons Learned for the Construction Inspection Program

The staff believes that this example shows the need to perform offsite inspections to support a determination by the staff that an ITAAC has been completed. As described in the main body of the framework document, the staff expects to perform inspections of the facility that fabricates the reactor pressure vessel pedestal. The inspections associated with this offsite fabrication facility would include but not be limited to the following:

–   a quality assurance inspection of the licensee to ensure that the details of the contract properly reflect the importance of this structure and clearly identify the quality control requirements for the structure

–   inspection of the offsite facility to ensure that the structure being properly manufactured, including a review of welding records and quality assurance and quality control

–   inspection of the onsite assembly of the modular structure, including a review of the processes to ensure that the structure was not damaged during shipping, assembly by the contractor, and final placement.

Appendix F
Generic Combined License From SECY-00-0092

[NAME OF NUCLEAR FACILITY]

[NAME OF NUCLEAR FACILITY OWNER]

Docket No. 52-[XXX]

License No. NPF-[XX]

1.   The Nuclear Regulatory Commission (the Commission) has found that:

A.   The application for a combined license (COL) filed by [name of nuclear facility owner(s) (the licensee)][, which references Appendix __ to 10 CFR Part 52,] complies with the standards and requirements of the Atomic Energy Act of 1954, as amended (the Act), and the applicable regulations set forth in 10 CFR Chapter I, and all required notifications to other agencies or bodies have been duly made;

B.   The applicable requirements set forth in 10 CFR 52.77, 52.78, 52.79, 52.81, 52.83, 52.85, 52.87, 52.89, [52.91, if applicable], and 52.97 [and Appendix __ to 10 CFR Part 52] have been met;

C.   There is reasonable assurance that the facility will be constructed and will operate in conformity with the application, as amended, the provisions of the Act, and the applicable regulations set forth in 10 CFR Chapter I, except as exempted from compliance in Section 2.F below;

D.   There is reasonable assurance that the activities authorized by this COL can be conducted without endangering the health and safety of the public and (ii) that such activities will be conducted in compliance with the applicable regulations set forth in 10 CFR Chapter I, except as exempted from compliance in Section 2.F below;

E.   The licensee is technically and financially qualified to engage in the activities authorized by this COL in accordance with the applicable regulations set forth in 10 CFR Chapter I;

F.   The licensee has satisfied the applicable provisions of 10 CFR Part 140, "Financial Protection Requirements and Indemnity Agreements."

G.   The issuance of this license will not be inimical to the common defense and security or to the health and safety of the public;

H.   The issuance of this license is in accordance with 10 CFR Part 51 and all applicable requirements have been satisfied; and

1.   The receipt, possession, and use of source, byproduct, and special nuclear material as authorized by this license will be in accordance with the applicable regulations in 10 CFR Parts 30, 40, and 70.

2.   On the basis of the foregoing findings regarding this facility, COL No. NPF-[XX] is hereby issued to [licensee], to read as follows:

A.   This license applies to the [Name of Nuclear Facility], a light-water nuclear reactor and associated equipment (the facility), owned by the licensee. The facility is located and is described in the licensee's final safety analysis report (FSAR), as supplemented and amended, and the licensee's environmental report, as supplemented and amended.

B.   Subject to the conditions and requirements incorporated herein, the Commission hereby licenses the licensee:

(I)   Pursuant to Sections 103 and 185.b of the Act and 10 CFR Part 52, to construct, possess, use, and operate the facility at the designated location in accordance with the procedures and limitations set forth in this license;

(2)   Pursuant to the Act and 10 CFR Part 70, to receive and possess at any time, special nuclear material as reactor fuel, in accordance with the limitations for storage and amounts required for reactor operation, described in the FSAR, as supplemented and amended;

(ii) Pursuant to the Act and 10 CFR Part 70, to use special nuclear material as reactor fuel, after the finding in Section 2.D(1) of this license has been made, in accordance with the limitations for storage and amounts required for reactor operation, and described in the FSAR, as supplemented and amended;

(3)   Pursuant to the Act and 10 CFR Parts 30, 40, and 70, to receive, possess, and use, at any time, any byproduct, source, and special nuclear material as sealed neutron sources for reactor startup, sealed sources for reactor instrumentation and radiation monitoring equipment calibration, and as fission detectors in amounts as required;

(4)   Pursuant to the Act and 10 CFR Parts 30, 40, and 70, to receive, possess, and use in amounts as required, any byproduct, source, or special nuclear material without restriction to chemical or physical form, for sample analysis or instrument calibration or associated with radioactive apparatus or components; and

(5)   Pursuant to the Act and 10 CFR Parts 30 and 70, to possess, but not separate, such byproduct and special nuclear materials as may be produced by the operation of the facility.

C. The license is subject to, and the licensee shall comply with, all applicable provisions of the Act, and the rules, regulations, and orders of the Commission, including the COL inspections, tests, analyses, and acceptance criteria (ITAAC) contained in Appendix A of this license.

D. The license is subject to, and the licensee shall comply with the conditions set forth in 10 CFR Chapter I, now or hereafter applicable [consistent with the requirements in Section VIII of Appendix __ to 10 CFR Part 52]; and the conditions specified and incorporated below:

(I) Nuclear Fuel Loading

The licensee shall state under oath or affirmation to the Commission that the acceptance criteria in the COL ITAAC have been met.

(ii) The licensee is authorized to load fuel into the reactor vessel and perform precritical testing (zero power) after the Commission has found, in accordance with 10 CFR 52.103(g), that the acceptance criteria have been met.

(2) Low-Power Testing

Upon approval of the Director of the Office of Nuclear Reactor Regulation, the licensee is authorized to perform low-power testing and operate the facility at reactor steady-state core power levels, not in excess of [XX] megawatts thermal (5-percent power), in accordance with the conditions specified herein.

(3) Maximum Power Level

Upon approval of the Director of the Office of Nuclear Reactor Regulation, the licensee is authorized to perform power ascension testing and operate the facility at reactor steady-state core power levels, not in excess of [XXXX] megawatts thermal (100 percent power), in accordance with the conditions specified herein.

(4) Incorporation

The COL ITAAC, plant-specific Technical Specifications, Environmental Protection Plan, and Antitrust Conditions contained in Appendices A, B, C, and D, respectively, of this license are hereby incorporated into this license.

E. The licensee shall report any violations of the requirements in Section 2.D of this license within 24 hours. Initial notification shall be made in accordance with the provisions of 10 CFR 50.72, with written follow up in accordance with the procedures described in 10 CFR 50.73.

F. The following exemptions are authorized by law and will not endanger life or property or the common defense and security. Certain special circumstances are present and these exemptions are otherwise in the public interest. Therefore, these exemptions are hereby granted.

[(1) LISTING OF EXEMPTIONS FROM DESIGN CERTIFICATION RULE (DCR)]
[(2) LISTING OF EXEMPTIONS WHICH ARE OUTSIDE THE SCOPE OF DCR]

G. The licensee shall fully implement and maintain in effect all provisions of the physical security, guard training and qualification, safeguards contingency plans, and all amendments made pursuant to the authority of 10 CFR 50.90, 50.54(p), 52.97[, and Section VIII of Appendix __ to Part 52] when nuclear fuel is first received onsite, and continuing until all nuclear fuel is permanently removed from the site.

H. The licensee shall have and maintain financial protection of such type and in such amounts as the Commission shall require in accordance with Section 170 of the Atomic Energy Act of 1954, as amended, to cover public liability claims.

I. The following operational requirements that are applicable to this license will become effective after the Commission finds that the acceptance criteria in this license (COL ITAAC) have been met in accordance with 10 CFR 52.103(g):

(1)   emergency plans,
(2)   technical specifications,
(3)   ....

J. After the Commission has made the finding required by 10 CFR 52.103(g), the COL ITAAC [not including the Tier 1 information from the referenced design certification rule (DCR)] do not constitute regulatory requirements either for licensees or for renewal of the license; except for specific ITAAC, which are the subject of a Section 103(a) hearing, their expiration will occur upon final Commission action in such proceeding.

K. This license is effective as of the date of issuance and shall expire at midnight on [the date 40 years from the date of issuance].

FOR THE NUCLEAR REGULATORY COMMISSION

Director
Office of Nuclear Reactor Regulation

Appendices:
Appendix A - COL ITAAC [including Tier 1 information]
Appendix B - Technical Specifications [plant-specific]
Appendix C - Environmental Protection Plan
Appendix D - Antitrust Conditions
Date of Issuance:

Appendix G
An example of an NRC Organization for Implementing
a Construction Inspection Program

The CIP team has proposed the following example organization to the CIP steering committee. The basic organization is similar to the Special Projects organizations of the late 1980s and early 1990s for TVA and Comanche Peak.

The CIP implementation may be performed using different organizations to address onsite construction and remote fabrication of systems, structures, and components. All activities could be coordinated through the onsite organization and be planned, scheduled, and tracked through completion using the CIP Information Management System (CIPIMS).

The basic construction inspection organization would exist on site. This organization would consist of six individuals, supported by regional and headquarters technical experts and inspectors. The organization structure is presented in Figure G-1 below.

The onsite organization would be lead by the site construction inspection supervisor (SCIS). This senior staff member (GG-15) would be responsible for all onsite NRC personnel and any associated activities involving NRC or NRC contract support personnel. The SCIS would report to the director of the division of reactor projects. The SCIS would serve as the staff inspection supervisor and senior resident inspector for much of the construction period, until the region determined that the site inspection activities and senior resident responsibilities needed to be separated. This individual would serve as the senior NRC staff member on site regardless of visitors or temporary assignees. All direct communication between the applicant and the regions or headquarters would be required to go through the SCIS. If a senior staff member for the applicant wanted to speak directly with the region, the individual would notify the SCIS prior to contacting the region. The SCIS would be directly responsible for all inspection activities performed on site.

The onsite inspection team would consist of a scheduler and three team leaders, one each for mechanical inspection activities, electrical and instrument and control (EIC) inspection activities, and civil and structural inspection activities. Miscellaneous inspection activities will be distributed amongst the team leaders by the SCIS. (See figure below.)

The scheduler would be trained in a scheduling program that is compatible with the applicant's scheduling software. This individual would also be trained on the use of the CIPIMS and would be responsible for the overall utilization and maintenance of the CIPIMS for that site. All efforts should be made to assign this individual for the duration of the project. The scheduler would be directly responsible for interfacing with the applicant's scheduling organization and coordinating all inspection activities with the onsite inspection team leaders. The scheduler would report directly to the SCIS.

The three team leaders would be responsible for all onsite inspection activities relating to his/her assigned discipline(s). Team leaders would be fully qualified with sufficient experience to perform and/or supervise the inspections within their purview. They would be responsible for ensuring that all resources are available and scheduled prior to scheduled inspection activities.

They would ensure that technical and/or inspection support personnel are properly briefed and appropriately prepared for the planned inspection prior to the start of any inspection activities.

Team leaders must be trained on the CIPIMS and would be ultimately responsible for developing and processing inspection reports, ensuring that the CIPIMS is properly updated, and preparing *Federal Register* notifications for onsite inspections, as appropriate. For any extended absences (in excess of one week), team leader coverage should come from outside the site organization. In addition to the onsite construction inspection team, a resident inspector would be assigned early on in the construction process. This individual would assume the routine resident responsibilities and interactions with the applicant, and would serve as a replacement for team members during routine absences and limited annual leave. This individual would be trained on the use of the CIPIMS, and become sufficiently familiar with scheduling activities to be able to replace the team scheduler for short periods. The resident inspector would be directly assigned to the SCIS until the region determined the need to assign a separate senior resident inspector.

Remote inspections or fabrication inspection/audit activities would be implemented by different implementing organizations that would coordinate their activities with the onsite construction inspection team using the CIPIMS. All remote inspection/audit activities would be coordinated by the director of the Division of Inspection Program Management. In general, inspection/audit activities for major U.S. manufacturers' would be assigned to the region responsible for the geographical area where the manufacturing is occurring. Foreign manufacturer inspections/audits generally would be performed out of headquarters. Special inspections/audits generally would be performed out of headquarters or "centers of excellence."

Centers of excellence are organizations, such as the regions, NRR, RES, and possibly NMSS, other Government agencies, or specialized consultants that have a concentration of specialized skills to perform inspections/audits of unique production activities. These production activities may involve software, monitoring instrumentation, fuel fabrication, safeguard components, etc. If multiple specialized skills are needed for a single inspection/audit, then a mix of organizations may be assigned to a single remote inspection/audit. If so, consideration needs to be given to the nature of the inspection/audit, and the organization that is providing the most support to the effort in determining the lead organization. The lead organization would be determined by senior executives representing each of the organizations involved. Difficulties in deciding the lead organization would be resolved by the Director, Division of Inspection Program Management. The lead organization would assign a team leader. The team leader would be responsible for the development of the audit/inspection report and updating the CIPIMS through the onsite organization. If headquarters were assigned as the lead organization, the appropriate project manager from the Division of Licensing Project Management would update the CIPIMS for the team. However, the ultimate responsibility for updating the CIPIMS would remain with the team leader. All remote audit/inspection reports will require SCIS concurrence.

All onsite and remote Inspection/audit information would be placed in the CIPIMS within 45 working days of completing the inspection. Updating of CIPIMS information would be completed within 15 working days of issuing the applicable inspection/audit report.

Nuclear Regulatory Commission Construction Inspection Program Organization Chart

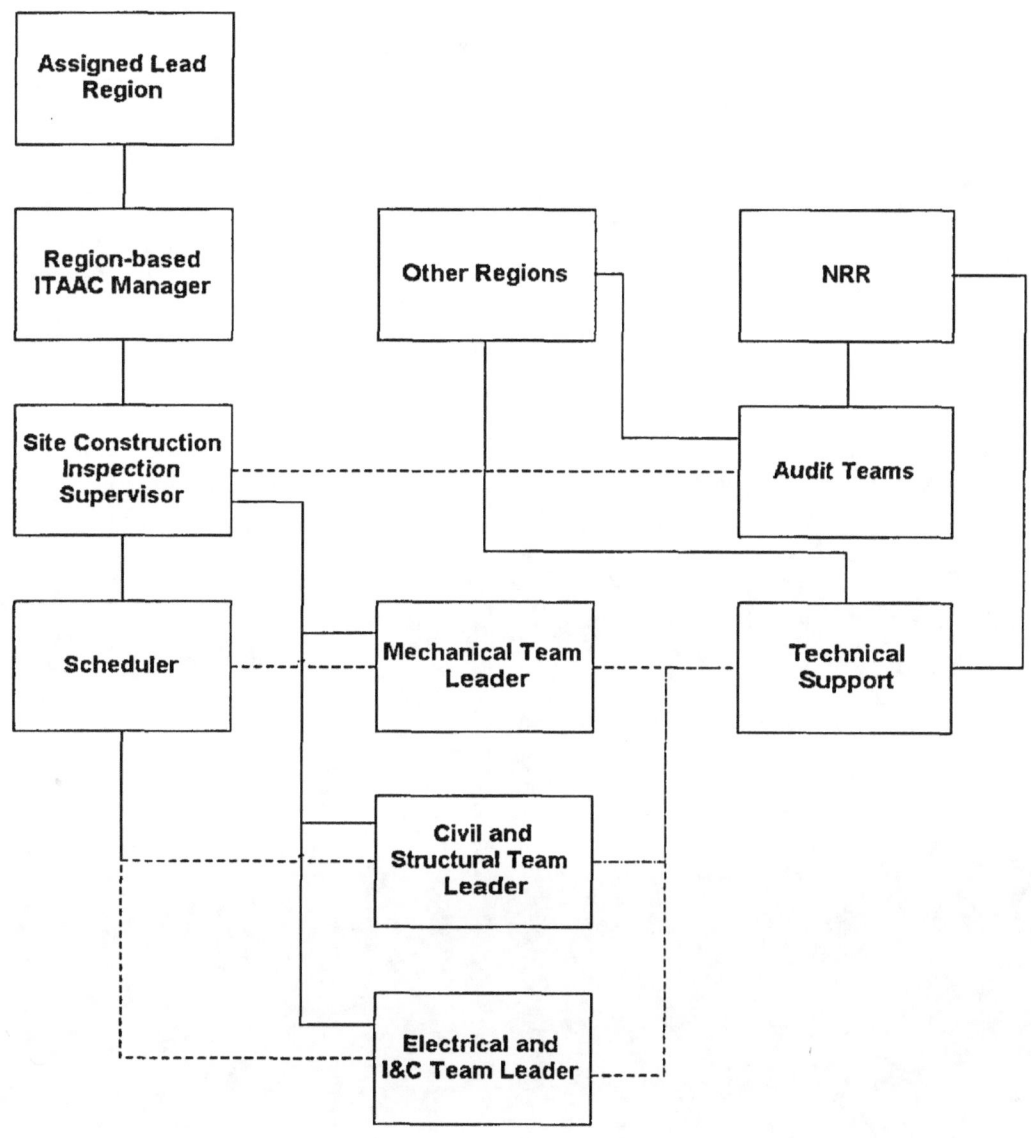

Appendix H
Comparison of AP600 and ABWR ITAAC to Inspection Procedures

SUMMARY

One objective of the Construction Inspection Program Team was to determine the scope of effort required to develop the procedures that will be necessary to verify that the ITAAC have been met. To meet this objective, the team obtained contract support from three experienced engineers. One had significant experience in construction and operational inspections, another had operational inspection experience, and the third had significant design experience.

The team tasked the contractors to review the ITAAC for the AP600 and ABWR designs and compare the acceptance criteria to the existing NRC inspection procedures. The purpose of this review and comparison was to identify any glaring holes that would require significant effort to develop procedures for guidance on inspecting and verifying the completion of ITAAC.

The result of the effort was encouraging. There were relatively few ITAAC that would require new procedures. Many of the existing procedures would require some revision to be fully capable of verifying the ITAAC, while some procedures were acceptable as written.

This appendix describes the contractors' methods and results. This provides a starting point for the development of the procedures that will be necessary to support the construction of new nuclear power plants under the regulations of 10 CFR Part 52. A future review will be performed to develop matrices for the two approved designs to identify all procedures that would be necessary for the verification of each ITAAC.

1.0 INTRODUCTION

Beckman & Associates, Inc. was contracted to evaluate the adequacy of NRC inspection procedures (IPs) in implementing the Construction Inspection Program (CIP) for new facilities. The CIP for facilities to be constructed in accordance with approved design control documents was evaluated by comparing the acceptance criteria listed in each design control document to the guidance available in IPs. The design control documents for the Westinghouse AP600 design and the General Electric ABWR design were reviewed.

The design control documents contained Tier 1 specifications for the systems, facilities, and programs that had been approved by the NRC. The Tier 1 documents included tables associated with the required Inspections, tests, analyses, and acceptance criteria (ITAAC). The acceptance criteria will require inspection verification as part of the CIP.

The three Beckman team members developed a spreadsheet for each of the two plant types that were evaluated. The spreadsheets summarize each of the acceptance criteria listed in the design control document, the IP determined to be applicable to inspect that acceptance criterion, and an evaluation of the adequacy of the guidance contained in the IP to fully inspect the acceptance criterion.

The Beckman team reviewed and evaluated IPs selected from all of the NRC inspection programs (Construction, Pre-Operational Testing and Operations Phases) that appeared to contain guidance that could be used to verify the adequacy of an acceptance criterion. Most of the IPs provided to the Beckman team members were from the NRC CIP team leader. Other procedures were obtained from the NRC electronic database, and some were found in personal hard copy files.

Some procedures listed in an older (November 1993) NRC Inspection Manual Index had titles suggesting that the procedures might contain applicable guidance, but copies of the IPs could not be located. Those IPs are mentioned under the applicable spreadsheet heading but were not credited for providing guidance.

The guidance provided in the IP selected for a given acceptance criterion was compared to the required verification activity to establish which areas had adequate inspection guidance and where additional guidance was needed. The adequacy of the IP to verify the acceptability of the acceptance criterion was then categorized using the following criteria:

Category 1:   There was essentially no IP that addressed the Acceptance Criterion.
Category 2:   The IP required major revision to fully address the Acceptance Criterion.
Category 3:   The IP required minor revisions to fully address the Acceptance Criterion.
Category 4:   The IP was essentially adequate to evaluate the Acceptance Criterion.

In assigning the category designation, Beckman team assumed that the construction verification activities would be completed during normal construction inspection activities. The construction verification would, in some cases, also fulfill the verification of completion of the acceptance criteria. The team tended to select a higher category designation for some IPs that did not provide detailed guidance for an attribute when the associated acceptance criteria required a specific verification activity.

Some of the same IPs were frequently found to contain inspection guidance applicable to more than one of the ITAAC areas. The category designation that was assigned to the IP in one area, or for one acceptance criterion within an area, was based on the particular requirement.

2.0 SUMMARY OF FINDINGS

The selected category designations listed in the spreadsheets were tabulated for each of the reviewed plants. When more than one category had been proposed by team members, the category was determined on a weighted basis considering the requirement and the guidance. The number of acceptance criteria listed in each of the categories is shown below:

DESIGN	CATEGORY			
	1	2	3	4
AP600	136	215	478	9
ABWR	99	115	975	233

These numbers show that 351 (42% of the total) acceptance criteria for the AP600 design and 214 (15% of the total) acceptance criteria for the ABWR design lack significant inspection guidance.

The details of the above findings were further evaluated for each design.

2.1 AP600 Summary

Sections 2 and 3 of the AP600 Tier 1 design control document presented the descriptions and ITAAC tables for the approved systems, facilities, and programs. There were 93 areas in Sections 2 and 3, but no ITAAC were associated with 37 of the areas. A total of 838 acceptance criteria was listed in the ITAAC tables for the remaining 56 areas.

No applicable IPs were discovered for three of the AP600 ITAAC Tables: 2.3.19, "Communications"; 3.2, "Human Factors Engineering"; and 3.7, "Reliability Assurance Program." Those sections of the design control document were also listed at the top of the spreadsheet. All of the other tables containing acceptance criteria were found to have at least partial IP guidance available for some of the listed requirements.

While few areas were considered to have complete, standalone guidance available, the IPs designated for over half of the total number of AP600 acceptance criteria were considered to be Category 3. For many of the items classified as Category 3, incorporating instructions in the designated IP to review the design control document and providing guidance to verify the acceptance criteria, could produce complete guidance.

There were 351 AP600 acceptance criteria that were determined to have little or no available IP guidance and were, therefore, judged to be Category 1 or Category 2. Although an applicable IP was designated for over half of these acceptance criteria (the 215 Category 2), the guidance provided in the available IP was considered to be lacking in detail or specificity and would require significant revisions and/or additions.

It should be noted that the number of Category 2 entries does not directly represent the number of IPs that require revision. The same IP was frequently listed as the applicable guidance for a number of different ITAAC requirements. The following IPs were listed as Category 2 guidance for numerous acceptance criteria:

IP	TITLE	# of Acceptance Criteria
37051	Verification of As-Builts	9
50100	Heating, Ventilating, and Air Conditioning Systems	7
70434	Engineered Safety Features Actuation System Test - Preoperational Test Witnessing	22
70444	Containment Isolation Valve Test - Preoperational Test Witnessing	9
93807	Systems Based Instrumentation and Control Inspection	17

The number of IPs that would require major revision to meet the Category 4 criteria is, therefore, considerably less than the total number of Category 2 acceptance criteria. There was not enough time to analyze IPs that were identified number of IPs in each category. It appears, however, that the number of Category 2 IPs requiring significant revision would be less than 35.

No applicable inspection guidance was located for 136 AP 600 acceptance criteria (16%). A review of the details for these category 1 acceptance criteria determined that many of the deficient areas could be grouped together and addressed in generic IPs. A listing of the generic IPs follows:

- Procedure for basic configuration/functional arrangement
- Procedure for review of certified stress reports
- Procedure for a hydrostatic/pressure test
- Procedure for mechanical separation of divisional equipment
- Procedure for motor-operated valves, check valves, power-operated valves, valve failure on loss of motive power
- Procedure for review of seismic analysis to withstand safe shutdown earthquake
- Procedure for review of structural analysis to withstand design basis accident loads
- Procedure for review of maintenance of containment integrity/isolation
- Procedure for review of ASME Design Report
- Procedure for review of ASME non-destructive examination (NDE) Report
- Procedure for review of Leak-Before-Break Report
- Equipment qualification procedure (mechanical and electrical)
- Procedure for physical separation and electrical isolation of test signals
- Procedure for test signal and displays, parameters, and controls in main control room

As examples of areas where similar requirements could be grouped together, 35 acceptance criteria required the review of ASME, NDE, and hydrostatic test results were designated Category 1 because specific IPs were not located. These areas could probably be inspected using generic IPs that provide guidance for that type of inspection. Likewise, a generic procedure for evaluating the environmental qualification of mechanical equipment could be developed to address the 11 mechanical environmental qualification acceptance criteria that were determined to be Category 1 because of the lack of an appropriate IP. Numerous other acceptance criteria required verification that a report was available to show the design value(s) had been met.

In addition, a final review of the AP600 spreadsheet found that 24 acceptance criteria required verification of specific details (e.g., tank volume, flow rate, or orifice size) that would require little additional inspection guidance. Specific items could be covered by a "generic" ITAAC inspection procedure that provided guidance on reviewing the design control document and verifying the acceptance criteria listed in tables.

Therefore, it would appear that minor changes to existing procedures and the development of a small number of generic IPs would decrease the number of AP600 Category 1 items from 136 to approximately 50.

2.2 ABWR Summary

The ABWR design control document was reviewed using the same methodology that had been used to review the AP600 document. The total number of ABWR tables and associated acceptance criteria was much larger than those provided for the AP600 design. There were 159 tables in Sections 2 and 3 of the ABWR design control document with a total of 1422 acceptance criteria listed. (No acceptance criteria were listed in 62 tables)

No applicable IPs were discovered for eight of the ITAAC tables in the ABWR design control document. Those ITAAC are listed below. All of the other ITAAC tables that listed acceptance criteria were found to have some amount of IP guidance available.

2.3.2   Area Radiation Monitoring Systems

2.11.11   Station Service Air System

2.11.12   Instrument Air System

2.11.13   High Pressure Nitrogen Gas Supply System

2.12.16   Communications

2.14.6   Atmospheric Control System (N2 Injection)

3.1   Human Factors Engineering

3.6   Design Reliability Assurance Program

The tabulation of the categorization of inspection guidance for the ABWR acceptance criteria is provided above. A large percentage (69%) of the items were considered to be Category 3 and, like the AP600 Category 3 items, many of those could be upgraded to Category 4 with minor revisions to existing procedures.

There were 214 ABWR acceptance criteria that were listed as Category 1 or Category 2 — 15% of the total number. While this number indicates that many new IPs are needed and major revisions to others are required, the number was much lower than for the AP600. The smaller number of items that the team determined had only Category 1 or 2 guidance available was considered to be mainly attributable to the number of ABWR acceptance criteria that involved specific verification actions (e.g., verify valve opens on signal, or verify opening time is less than 3 seconds). The team's familiarity with the IPs may have also led to the somewhat higher ratings of the procedures.

The ABWR spreadsheet was also found to contain a number of IPs that had been determined to provide only Category 2 guidance for a number of different ITAAC requirements. The following IPs are examples:

IP	TITLE	# of inspection criteria
70432	Control Rod System Test - Preoperational Test Witnessing	17
93807	Systems Based Instrumentation and Control Inspection	15

In addition, the same IPs that were designated Category 2 for some attributes were noted to be rated as providing Category 3 or 4 guidance for others. The number of IPs requiring major revision to provide the Category 4 guidance for ABWR acceptance criteria was, therefore, also considered to be less than the total number of Category 2 attributes.

The ABWR acceptance criteria were reviewed again to determine areas where similar requirements could be grouped together so that inspections could be conducted using general guidance. The findings were similar to the AP600 findings. For instance, there were nine requirements to verify specific values that could be inspected using a generic ITAAC verification IP. There were also 20 acceptance criteria for system hydrostatic/pressure test results verification that could be inspected using one or two general leak test IPs.

Based on an overview of the ABWR acceptance criteria designated as Category 1 and Category 2, it appears that the number of IPs that need to be developed and/or significantly revised is around 60.

3.0 CONCLUSIONS

There were 351 AP600 and 214 ABWR acceptance criteria identified as having insufficient inspection guidance provided by existing IPs. Those acceptance criteria were designated as Category 1 and Category 2 in the applicable spreadsheet. As discussed above, however, the number of acceptance criteria lacking sufficient inspection guidance does not directly correspond to the number of IP changes that would be needed. The same IP was frequently found to provide varying levels of guidance for different acceptance criteria. Therefore, the revision of one IP could improve the rating of a number of Category 1, 2 and 3 entries.

In addition, many of the same IPs were found to be applicable to both plant designs that were reviewed. Therefore, one revision could affect the category rating of a given IP in more than one location, in more than one plant design. Constraints did not allow a determination of the number of these "common" IPs.

The total number of new IPs that need to be developed and existing IPs that require major revision was considered to be around 100.

Most of the existing IPs will require at least slight revisions to include guidance for reviewing the applicable section(s) of the design control document and to verify completion of any related acceptance criteria.

An evaluation of the total number of IPs that were identified in the spreadsheets and an analysis of the necessary additions would help to determine the total level of effort required to produce acceptable levels of inspection guidance for all of the acceptance criteria.

Appendix I
Comment Resolution Summary for the
Draft Construction Inspection Program Framework Document

NOTE: Comments followed by an item number refer to items in the letter and its attachments (ML 033090096) provided by NEI and endorsed by a letter from Southern Company (ML033090101).

Comments followed by a page number are comments/questions from the transcript of the August 25 2003 public workshop on the framework document which the NRC committed to treated as public comments. (ML032790347)

Comments followed by (AC) refer to recommendations from the Advisory Committee on Reactor Safeguards (ML033510735)

Comment / Recommendation/ Question	Resolution
Topic: IMC-2501 - ESP	
1 How does the NRC intend to ensure that contractors used by the NRC are trained to the new standards etc? (Page 47)	Contractors employed by the NRC for either headquarter's staff review of an ESP application or participation in a Region lead inspection will receive orientation training before beginning work. Additionally, such contractors will receive on-going direction by the NRC ESP project manager or Regional inspection team leader. However, no change was made to the document because this is at a greater level of detail than intended for the Framework Document.
2 Will Part 21 be applied to an applicant? (Page 42)	New, Research and Test Reactors (RNRP) staff and the Office of the General Counsel (OGC) are aware of the issues surrounding the applicability of Part 21 to ESP applicants and are reviewing the matter.
3 In IP 35002, for Part 21, what is the applicability to contractors? (Page 38)	
4 Reconsider the applicability of Part 21 to ESP applicants. (Letter item 1)	
5 Modify IP 35002 to eliminate the reference to Part 21 applicability to ESP applicants. (Item A1.2)	

	Comment / Recommendation/ Question	Resolution
6	Revise paragraph B2 on page 7 to reflect that an ESP provides approval of a site for one or more plants and may not expire when a COL or CP is issued. (Item A1.3)	The paragraph has been revised to remove the misleading statement.
7	What are the expectations regarding Appendix B? (Page 32)	The Emergency Preparedness and Plant Support Branch issued a letter to NEI on August 4, 2003 which stated that the NRC expects that an ESP applicant will use a QA control framework equivalent in substance to that described in Appendix B. Review Standard (RS) -002 provides one method for demonstrating that the ESP QA controls are equivalent in substance to Appendix B. However, an applicant's failure to use a framework equivalent to Appendix B would not, in and of itself, result in rejection of the application. Further, any deficiencies or deviations in an applicant's quality assurance measures would need to be addressed to ensure the reliability and integrity of data contained in or supporting the ESP application. The staff believes that Section 17.1.1 of the review standard allows sufficient flexibility for an applicant to propose alternate QA measures, and that no change to the text is necessary as a result of these comments.
8	Is a deviation or deficiency in accordance with Appendix B something that the applicant would have to address?(Page 51)	
9	Will the NRC staff request a QA program through other means? (Page 54)	
10	Modify the framework document to clarify the expectation that ESP QA measures be equivalent in substance to Appendix B. Clarify also in IP 35002 and Review Standard-002. (Item A1.1)	
	Topic: Construction Inspection Program Information Management System (CIPIMS	
11	How are the licensees going to share construction schedules if the information contains proprietary details? (Page 92)	Meeting(s) will be scheduled in the future to explore further how information could be shared.
12	How and how often is the NRC going to get information from the licensee and vendors to support CIPIMS? (Page 109)	

Comment / Recommendation/ Question	Resolution
13 Schedule follow-up interaction to discuss coordination of construction and inspection schedules, use of coding schema, protection of proprietary information and business sensitive schedules. (Letter item 2 and Items A3.1.1, A3.1.2, A3.1.3)	
Topic: IMC-2502 - Pre-COL Phase	
14 Clearly identify that a principal objective of NRC EDV is to provide reasonable assurance that detailed design information on which construction will be based is consistent with the design approved during a design certification or COL review. (Item A2.1.1)	No change is needed since the EDV inspection objective stated in the framework document quotes SECY-94-294. Verifying that an applicant's design is consistent with the approved design, during either design certification or COL review, is one purpose of the EDV.
15 Modify the framework document to reflect that the engineering design verification (EDV) need not be completed prior to COL issuance. (Item A 2.1.2)	SECY-94-294 indicated that as plant construction progresses, NRC will determine if the engineering design is adequate. The NRC will use performance-based inspections to verify that plant systems and components are installed and tested to applicable standards, certified design information, and ITAACs. Thus, the EDV inspections would continue during construction after the COL application had been approved. The wording in the framework document was revised to reflect that the staff plans to conduct these inspections as early in the process as practical but that they may continue after the COL is issued.
16 Modify the framework document to reflect that the scope of EDV may encompass review of additional topical design areas such as fire protection, environmental qualification, seismic design, HELB analyses, and separation/independence. (Item A2.1.3.1)	No change made. The framework document will not provide an all inclusive list of the different inspection areas reviewed by the EDV inspections. Design areas like those referenced are unique to the applicant's specific design and their associated ITAACs would denote the design requirements to be met on a per system basis. Topical reports submitted by the licensee would be reviewed during the COL application stage.

	Comment / Recommendation/ Question	Resolution
17	Include public notification via a Federal Register notice or other method of the NRC determination that the licensee design engineering processes are acceptable. (Item 2.1.3.2)	No change made. The NRC will use inspection reports as the primary vehicle to inform the licensee and the public about the result of inspection efforts, including the results of the NRC's review of the design engineering process.
18	Include information stating that after the NRC has approved the design engineering process, follow-up inspection to spot check the process would focus on configuration management and design details completed after the main thrust of NRC engineering design verification was completed and had established the acceptability of the licensee's overall design engineering processes. (Item A2.1.3.4)	The information about the design engineering process has been revised to reflect that the NRC does not approve the applicant's design engineering process, but rather assesses its viability. The NRC inspections will determine whether the applicant abides by that process in actuality. The NRC will adjust its inspection effort either up or down based on its determination of whether or not the design engineering process is working as expected.
19	Incorporate the idea that the main thrust of NRC engineering design verification would focus on design areas other than those covered by DAC (e.g., piping, instrumentation and control, and the main control room), unless the applicant chose to complete and seek NRC approval in the COL of all or a portion of the plant design in such DAC areas. After staff reviews in areas with DAC are complete, EDV in these areas may be accomplished as a follow-up to the main EDV milestone achieved at the time of COL issuance or early in construction. Or, perhaps more likely, the staff safety reviews and EDV may occur in parallel as the plant design in DAC areas is completed. (Item A 2.1.3.4)	No change made. Areas currently covered by DAC, such as those described in SECY-94-294, where design descriptions and functional system drawings are adequate for licensing reviews but not for actual construction or construction inspection activities, would be appropriate for inspection. The timing of the reviews of DAC will depend on when the applicant actually provides sufficient information to warrant the inspections.

	Comment / Recommendation/ Question	Resolution
20	Clarify that it is at least possible and perfectly acceptable under Part 52 that a COL applicant might not contract for major components, detailed design engineering or construction until after a COL is issued. (Item A2.2)	Document revised to reflect that rather than waiting until the end of construction, the NRC will start its review as soon as it can based on the applicant's readiness.
21	Distinguish between licensing reviews of operational program descriptions based on the SRP or other COL application review guidance versus operational program readiness inspections prior to plant operation in accordance with IMC-2504. This distinction should be made in the Framework document regardless of the outcome of parallel interactions concerning the extent of operational program information to be provided in COL applications. Regardless of the outcome of those interactions, the Framework document should reflect the focus of IMC-2504 on inspections to determine operational program readiness prior to operation. (Item A2.3)	The document has been revised to identify that the operational programs will receive two evaluations. The first will be a review of the program description as part of the COL application. The second evaluation will be an inspection that will take place prior to plant operation and will focus on the licensee's readiness to implement the program.

Comment / Recommendation/ Question	Resolution
Topic: IMC-2503 - ITAAC Verification	
22 If an inspection of ITAAC activities indicates that there are deficiencies that have not been addressed by the licensee's corrective action program, then the licensee should be required to identify and correct the weakness in the correction action program that lead to the deficiency. (AC)	The section related to ITAAC determinations has been revised to incorporate this recommendation.
23 SAYGO ITAAC - What does that mean? (Page 140) What does 'sign' mean in sign as you go? (Page 156)	The use of a SAYGO approach was designed to help the staff support the Commission in their determination that the ITAAC have been met. Because the existing ITAAC are based on complete systems being installed and do not recognize the multiple stages of construction, the use of SAYGO is envisioned as a tool by which the staff can signify overall satisfaction with the licensee activities that have been completed on an ITAAC to that point leading up to the
24 The staff is creating ITAAC where there is not really an ITAAC (Page 151)	completion of an ITAAC. The staff has selected that approach because it will allow for an on-going assessment to determine if the staff is on track to complete the necessary and sufficient reviews to support the 10 CFR Part 52.103(g) finding by the Commission. The description of SAYGO in the framework was revised to incorporate the key points of this explanation.
25 What would a 52.99 notice look like? (Page 159)	No change made. The Framework document, by design, will not contain detailed information, rather it acknowledges what will exist within the overall program. Specific information on form and format of NRC documentation will be covered in implementing procedures including manual chapters and inspection procedures.

	Comment / Recommendation/ Question	Resolution
26	How would the staff go about making the recommendation to the office director or regional administrator? Would it be mechanical or is it gathering additional information to supplement or a whole new assessment? (Page 184)	The document was revised to reflect that the recommendation to the NRR office director or regional administrator will be based on a review of the information associated with ITAAC including assuring that a determination letter was received, reviewed, accepted, and has been noticed in the Federal Register by the NRC.
27	Modify the framework document to reflect that SAYGO process conclusions provide confidence in the acceptability of quality-related construction processes, including conformance with applicable codes and standards, QA Program requirements, etc. (Item A 3.2.1)	Language revised to eliminate the use of 'conclusion'. The section now shows that programs and processes are part of the evaluation.
28	Expand the list of example processes identified on p. 17 of the framework document to include still other construction-related processes that may be amenable to early, systematic assessment and determination of acceptability by the NRC, such as receipt inspection, commercial grade dedication, warehousing and others. (Item A 3.2.1)	No change made. The table on page 17 is not intended to be all inclusive.
29	Either clarify that SAYGO ITAAC conclusions are SAYGO process conclusions that correspond directly to ITAAC acceptance criteria or the concept of SAYGO ITAAC conclusions could be eliminated. (Letter item 5 and Item A3.2.2)	See comments to 23 and 24.
30	The distinction between ITAAC conclusions by the NRC staff and the Commission's ITAAC finding is clear without the word "interim." Consider using the term "Section 52.99 ITAAC conclusions," (Letter item 3 and Item A3.2.3)	Change made. The term "ITAAC conclusion" has been changed to "ITAAC determination."

Comment / Recommendation/ Question	Resolution
31 Expand the discussion regarding independent review of ITAAC verifications to include clarifications that the independent review of the ITAAC completion would not involve re-review of all ITAAC but rather would be a 'vertical slice audit' and could begin in advance of fuel load and in parallel with ITAAC verification activities in the region. (Item A3.3.1)	The document has been revised to indicate that the NRC intends to complete a review of the documentation presented by the licensee in the ITAAC determination letter as it is received. This review will ensure that the NRC has considered every ITAAC. The staff will also review any inspection information related to each specific ITAAC when it is submitted by the licensee as complete.

When the licensee has informed the NRC that they have completed all of the ITAAC, a final independent review will consist of an audit that will ensure that the NRC has received a determination letter for each ITAAC, agrees that it is complete, and has published the required Federal Register notification in accordance with 52.99. |
| 32 Define and incorporate the process for triggering the Section 52.103(a) notice consistent with the description from the November 2001 NEI white paper. (Page 189, Letter item 6 and Item A 3.3.2) | The information was incorporated into the document. The added information reflects the NRC's November 20, 2003 response to NEI on their November 2001 white paper. (ADAMS Accession No. ML032760053) |
| 33 Include a statement that NRC ITAAC documentation, including Section 52.99 notices, should focus on the licensee's ITAAC determination bases. Matters not material to ITAAC determinations would be the subject of normal NRC inspections and reports. (Letter item 6 and Item A3.3.3) | The document has been revised to indicate that the "normal NRC inspections" are the means by which NRC will establish confidence in the licensee's construction program. In addition to a completeness review of the documentation submitted by the licensee with the ITAAC determination, inspection results related to that ITAAC will establish the bases for NRC acceptance of the licensee's ITAAC determination bases. |

Comment / Recommendation/ Question	Resolution
34 Revise the information related to rescinding a prior ITAAC conclusion to reflect an approach that would rely on the licensee's corrective action program to address most issues affecting installed system, structures or components that arise after an ITAAC is complete and a 52.99 notice is issued. (Item A 3.3.4)	The licensee's corrective action program will be an integral part of addressing inspection issues. The framework document has been revised to reflect that the NRC envisions that in most cases, items identified by the NRC will be turned over to the licensee to be addressed through the corrective action program. However, if NRC identifies new and significant information that calls a previous ITAAC determination into question, we will consider rescinding it. This decision would not be taken lightly, and would be a deliberate NRC management decision. Although the licensee may use their corrective action program processes to address the reason for rescinding any ITAAC determination, the nature of this decision will call for the NRC to closely monitor how the licensee resolves the issue.
35 The staff should reserve public meetings to exchange information regarding ITAAC deficiencies for situations when there are particularly significant negative findings necessitating involvement of NRC and licensee senior management. (Item A 3.3.5)	Agree, no revision to the document needed.
Topic: IMC-2504 - Non-ITAAC Inspections	
36 The Framework document should be modified to reflect that IMC-2504 will begin after the COL is issued. IMC-2504 should focus on (1) non-ITAAC inspections prior to fuel load (primarily ORAT inspections) that will support Region and NRR recommendations regarding readiness to load fuel, and (2) post-fuel load inspections prior to power operations (primarily start-up testing inspections). (Item A4.1)	The title of IMC-2504 has been revised from "Preparation for Operations" to "NON-ITAAC Inspections" to better reflect the the range of inspection activities to be covered.

The scope of inspections to be conducted under the new title will include the inspection activities to begin before the COL is issued. |

	Comment / Recommendation/ Question	Resolution
37	The gap between the ITAAC completion and the point at which tech spec surveillance requirements become effective should not be referred to as the regulatory gap since the licensee would be implementing QA, design control, work control, configuration control during that time. (Page 200)	Document revised to remove "regulatory gap"
38	On p. 24 of the Framework document, the staff uses the term regulatory "gap" to describe the time between when an individual ITAAC is complete and when the Commission makes it's Section 52.103(g) finding and discusses the need for inspections to ensure that the licensee is "managing this 'gap' appropriately. While such inspections may be appropriate, it is incorrect and misleading to refer to a regulatory"gap," and we recommended that the staff use different terminology to describe these inspections. (Item A 4.2)	

Comment / Recommendation/ Question	Resolution
Topic: Specific Comments on the Draft Framework Document	
39 On page 5 clarify (1) that ITAAC can be shown to be complete only after the underlying construction, inspection and test activities are complete. This necessarily means that demonstration of ITAAC completion will occur later in construction for some ITAAC versus others. And (2) ITAAC verification by the NRC will be based on SAYGO and other NRC inspection conclusions that are material to the ITAAC conclusion. (Item B.1)	Document was revised to clarify how SAYGO will be used.
40 In the discussion of IMC-2502, clearly state that the control room ITAAC and other "design acceptance criteria" are not required to be completed at time of COL issuance. (Item B.2)	Because design acceptance criteria are ITAAC, the only requirement is that they be completed prior to fuel load. If designs are not complete before the COL is issued, then they will be carefully reviewed as they are completed. A provisions for design inspections will be included in IMC-2504. Document revised to indicate that these ITAAC could be completed prior to COL.
41 Correct reference on page 11 in section D.1 to read Section 52.79(b)(1). (Item B.3)	No change made. Reference is correct as written in the framework document.
42 If ITAAC information is used as an example the acceptance criteria should be stated verbatim. (Item B.4.)	Agreed
43 Modify the statement on page 19 to read as follows: "Upon receipt of an ITAAC determination letter, the NRC staff will base its decision regarding ITAAC acceptability on a review of the licensee's ITAAC determination record and/or on NRC inspection reports and NRC SAYGO documentation that are material to the ITAAC in question." (Item B.5.)	Changes have been made to the statement to reflect the intent of this comment.

	Comment / Recommendation/ Question	Resolution
44	On p. 19, the Framework Document says "the staff will perform an independent review to ensure that it has received an ITAAC determination letter for each ITAAC and the staff agrees that all the ITAAC have been met." This language should be modified to reflect the purpose as clarified by the NRC staff during the August 27 workshop to audit and independently verify the ITAAC verification activities of the primary regional inspectors. As discussed in comment A.3.3.1, above, a 100% re-verification is not envisioned; it is expected that sampling and vertical slice audit methods would be used by the independent review team. (Item B.6.)	The document has been revised to reflect at a minimum the NRC will review the available information to ensure that the agency has received, reviewed, accepted, and published a notice in the Federal Register for each and every ITAAC. In addition, a sample of the ITAAC packages may be selected for review as a further assurance of the accuracy of the data.
45	On page 23 of the Framework Document, we recommend this statement be modified as follows pending the final resolution of the programmatic ITAAC issue: "~~Therefore, if~~ Regardless of whether or not an operational program ~~does not have~~ has an ITAAC, there is an expectation that the staff will perform inspections prior to operation to verify the licensee's compliance with regulations." (Item B.7.)	Document revised to incorporate the intent of the comment but not the exact wording stated.

Comment / Recommendation/ Question	Resolution
46 By definition, operational program inspections under IMC-2504 are separate from ITAAC verifications under IMC-2503. Therefore, on page 23 of the Framework Document, we recommend this statement be modified as follows: "To the extent these [transition to ROP] inspections are performed prior to loading fuel, these inspections will also supplement the bases for the regional administrator's recommendation to the Director of NRR regarding ~~ITAAC~~ plant readiness to load fuel." This change is consistent with language on p. 24 regarding consideration of ORAT results. (Item B.8.)	Document revised to incorporate the intent of the comment but not the exact wording stated.
47 On page 24, the Framework Document states that programs such as technical specifications must be in place and fully functional prior to the 52.103(g) finding. This statement is not accurate and conflicts with an earlier statement on the same page, which states that technical specifications will not become effective until the NRC issues its 52.103(g) finding. This page should be revised to indicate that the licensee must be ready to implement the technical specifications and other applicable operational programs prior to the 52.103(g), and not that they be "fully functional" before that finding. (Item B.9.)	No change was made because the document states that - "Prior to the Commission findings, the staff expects the licensee to phase in programs such as technical specification controls so that problems are recognized and solved before the program is required (by regulations or license condition) to be fully implemented."

	Comment / Recommendation/ Question	Resolution
48	On p. 25 of the Framework Document, the staff envisions separate NRC authorizations after the Commission makes its 52.103(g) finding to go above 5% power and to full power. The staff notes that the Commission approved these authorizations in the SRM on SECY-00-0092. However, the staff recommendation and the Commission approval of separate low- and full-power authorizations occurred without discussion with stakeholders of whether these actions are consistent with Part 52 and before the impact of these actions could be fully explored. (Item B.10.)	The document was modified to reflect the staff's current understanding that the COL will contain license conditions. The document reflects how recommendations will be made if conditions are part of the license. The underlying issue of whether or not a COL should include conditions and what the form of any such conditions should be will not be resolved in the framework and has been turned over to the New Reactor Licensing Section.
49	A target, such as 30-days from receipt of an ITAAC determination letter from the licensee, should be established for NRC to complete the ITAAC verification process and issue the required 52.99 notice. (Item B.11)	No change made. Assigning a target for completion is premature. However, the staff should establish due dates for completion of the work as it is received for review. Assigning due dates will ensure that the work is timely but also considers the overall volume of received items . The actual process to be used for reviewing, accepting and noticing an ITAAC determination package will be detailed in an NRC implementing procedure such as a manual chapter.
	Topic: Other comments and questions	
50	Will the Inspection Manual Chapters be issued for public comment? (Page 215)	The Inspection Manual Chapters will not be issued for public comment. The Inspection Manual Chapters will reflect the approach outlined in the final Construction Inspection Program Framework Document, on which the public was provided the opportunity to comment.

NRC FORM 335 (2-89) NRCM 1102, 3201, 3202	U.S. NUCLEAR REGULATORY COMMISSION	1. REPORT NUMBER (Assigned by NRC, Add Vol., Supp., Rev., and Addendum Numbers, if any.)
	BIBLIOGRAPHIC DATA SHEET *(See instructions on the reverse)*	NUREG-1789

2. TITLE AND SUBTITLE

10 CFR Part 52 Construction Inspection Program Framework Document

3. DATE REPORT PUBLISHED	
MONTH	YEAR
April	2004

4. FIN OR GRANT NUMBER

5. AUTHOR(S)

6. TYPE OF REPORT

7. PERIOD COVERED *(Inclusive Dates)*

8. PERFORMING ORGANIZATION - NAME AND ADDRESS *(If NRC, provide Division, Office or Region, U.S. Nuclear Regulatory Commission, and mailing address; if contractor, provide name and mailing address.)*

U.S. Nuclear Regulatory Commission

9. SPONSORING ORGANIZATION - NAME AND ADDRESS *(If NRC, type "Same as above"; if contractor, provide NRC Division, Office or Region, U.S. Nuclear Regulatory Commission, and mailing address.)*

Office of Nuclear Reactor Regulation
U.S. Nuclear Regulatory Commission
Washington, DC 20555-001

10. SUPPLEMENTARY NOTES

11. ABSTRACT *(200 words or less)*

The Information contained in the Construction Inspection Program Framework Document, NUREG-1789, details the overall philosophy and approach that will be used to inspect new nuclear power plants being licensed and built under 10 CFR Part 52. The information contained in this NUREG will guide the development of Inspection Manual Chapters and Inspection Procedures that will be used to implement the construction inspection program.

12 KEY WORDS/DESCRIPTORS *(List words or phrases that will assist researchers in locating the report.)*

inspection
construction
licensing
nuclear power plant

13 AVAILABILITY STATEMENT

unlimited

14 SECURITY CLASSIFICATION

(This Page)

unclassified

(This Report)

unclassified

15. NUMBER OF PAGES

16. PRICE

This form was electronically produced by Elite Federal Forms, Inc.

FRAMEWORK DOCUMENT